UKRAINE—STARTING A WIRELESS BUSINESS BY TWO GUYS FROM CLEVELAND

Myron Stoll

Published internationally by Myron Stoll
Cleveland, Ohio

Table of Contents

Glossary

Walter Bazarko. Part owner of Communications Technologies of Ukraine. Member of the Boards of Hatwave and Ukrainian Wave

Olha Bilichenko. Second Managing Director of Ukrainian Wave

EBRD. European Bank for Reconstruction and Development located in London. Provided the debt capital to Ukrainian Wave

Hatwave. The Western owned company which owned 48% of Ukrainian Wave.

HW. The shareholders were Communications Technologies of Ukraine, Hughes, the Manufacture of the telephone equipment, and OTE, the Greek telephone company

Hughes. Hughes Network Services, HNS, was the manufacturer of the wireless telephone.

HNS. Equipment and an investor in Hatwave

Valerij Kalynyuk. Formerly a Deputy Director of the L'viv Oblast, and the first Co-Managing Director of Ukrainian Wave

Hryhoriy Kirpa. Chief of the Railroad for Western Ukraine. Later became the Minister of Transportation for Ukraine. Member of the Board of Directors for Ukrainian Wave.

L'viv. The largest city (pop. 800,000) in Western Ukraine.

The city where Ukrainian Wave established its business.

Jerry Merkelo. First Co-Managing Director of Ukrainian Wave

OTE. The Greek Telephone Company located in Athens Greece. The State owned company is one of the largest companies in Greece. OTE was the largest shareholder of Hatwave

Bill Schlageter. Co-founder of Communications Technologies. Chairman of the Board of both Ukrainian Wave and Hatwave

Arunas Slekeys. Vice-President of Hughes, the manufacture of the telephone equipment, and a Member of the Board of Ukrainian Wave and Hatwave

Myron Stoll. Co-founder of Communications Technologies, Executive Vice-President of Ukrainian Wave, and a member of the Board of both Ukrainian Wave and Hatwave

Ukraine. Second largest country in Europe (Russia the largest). Borders: Moldova, Romania, Hungary, Czechoslovakia, -the Carpathian Mountains-Poland, Belarus, Russia—the Black Sea.

Ukrainian Wave. The Company established to provide fixed wireless telephone service UW to the city of L'viv. The company was owned on the Ukrainian side by the Ukrainian Government, the Railroad, and a small Ukrainian business man. On the Western side was HW.

Preface

The preface to this book is an email received from Karl Herold, now the retired partner of the law firm Jones Day in charge of the Frankfurt office, and the lawyer in charge of our project and two paragraphs from Lute Harmon Sr.'s editorial in the July, 2009 issue of "Inside Business." Both kindly read the manuscript, and both kindly made these comments.

Dear Kathy and Myron,

I finally dared to hit the "Reply" button, but what to say? I have read many books, but lived in none of them. In yours, I find cause to reflect on many great moments (like playing chess in the train to L'viv).

This endeavor to take a dream through its various stages of hopes, realizations, set-backs, and successes, both minor and devastating, without ever giving up, is indeed the fulfillment of that very dream. It reminds me of the story in "The Alchemist" by Paulo Coelho. Here the short version of it: A boy goes to the wise man to get an answer to his question as to the true meaning of life. The wise man tells him to take this spoon filled to the top with olive oil and walk through the wise man's garden and return in two hours without having spilled any of the oil. The boy returns in two hours and has not lost one drop of oil. The wise man then asked the boy what he had seen in the garden, had he seen certain flowers and trees? The boy responded that he saw none as he only concentrated not to spill any oil as instructed. The wise man was disappointed as the boy had accomplished the task without seeing or knowing the beauty of the garden. So he sent him again into the garden. This time the boy came back having seen all the flowers and the trees, fascinated by what he had seen. I cannot remember if in the story the boy had spilled all the oil on his second trip through the garden. It is not relevant anyhow. The secret of life is in the process and you have seen that process as you, with unbending determination, wondered through the garden of the deepest friendships, the climb of the highest personal mountain, the agony of disappointment, the breach of trust, the fragile nature of life and devotion, observation of hunger for power and money, the unbending commitment to principle, the pleasure of realizing the worth in others no matter their appearances and shortcomings, the willingness of cooperation and the desire to experience pure joy.

Myron, for me it was an honor to have been a part of your and Bill's team and there is no question that the oil in the spoon was not the accomplishment of a wireless network in L'viv, but the privilege of having accompanied you, albeit for only a very short span of our lives. Every minute so spent brought joy and true value into my life. For that I am grateful and you will always be in my thoughts. With the best of greetings to you and Kathy.

Karl

Two paragraphs from Lute Harmon Sr.'s editorial review, "A Great Story You'll Never Read"

> "Myron, I said, I have four observations—three you are going to like, and one you're not. The book reads like a who done it mystery, and I couldn't put it down, it is one of the most inspiring and moving stories I've ever read, it should be required reading for every entrepreneur in America; it will never get published."

> "If the Two Guys from Cleveland never gets published, the saddest fact will be that people will not get the opportunity to see the heart of a true entrepreneur, for it is two stories, not one. The most important story is a spiritual journey of two entrepreneurs who pursued a dream, never lost faith, developed deep friendships in a foreign country, maintained the highest standards of integrity when the obvious choice was not to and, if asked if they would do it again, would answer with a resounding "Yes.""

Most people ask why?
The dreamer asks why not?

Part I: Introduction

Why Am I Telling This Story?

This is a story about two guys from Cleveland, Ohio, Bill Schlageter and the author Myron Stoll who tried to do something we had never done before, start a business. We did this in a country we knew nothing about, had never visited, and did not know the language—even the letters of that language seemed strange. But the key in doing this was not doing something totally new, but building on the knowledge and skills we had acquired. It is the story of how we started a wireless telephone company in L'viv in Western Ukraine and what happened in the years afterward. It is a story about a journey that lasted sixteen years, beginning one year after Ukraine gained its independence. The purpose of the journey was to start a business, but the people on that journey created a complex story.

The purpose in writing this story is to share this journey, with the hope that this story will encourage, not discourage, people to apply their skills to do something they want to do, but have never done. To do this is a real challenge because the endless variables of life make it difficult to predict success. It takes an abundance of enthusiasm and desire to sustain the hard effort. It is people who made the difference. It will become a privilege and wisdom will be gained in walking through and seeing the new garden in your life.

Despite the substantial difficulties we encountered, Ukraine, like so many countries around the world, was a good place to do business, especially for the American entrepreneur. Ukraine, like many countries around the world is making the effort to modernize its economy. Ukraine not only has many natural resources and an industrial base, its people are well-educated, smart, and hard-working. They want to learn how to build successful businesses. The engineers, marketing, and technical employees in our company were all outstanding. There were great opportunities to add skills they did not have and build on what they did have.

We saw the critical importance of American values, honesty, hard-work and flexibility and working with diversity. This plays into the strengths of the American entrepreneur. These values need not and must not be abandoned.

We saw the importance of the bottom-up approach, not the top down. Go to the city or region where one intends to do business and build support from within. This not only creates more opportunities, it plays directly to the strengths and capabilities of the entrepreneur.

The essence of the American economy, its backbone, is the entrepreneurs who successfully start small and medium size business. Some grow to be large international businesses. The growth of the global economy, like the growth of the American economy, is not just for the large multi-national companies. In fact their size and their corporate requirements may not allow them to adapt to the ways of another society.

It is also a story about a country, region, and city which most of us know nothing about. Ukraine is the second largest country in Europe, second only to Russia. Western Ukraine is a region of the country which has been occupied by Russia, Poland, Lithuania, the Ottoman Empire, and the Austro Hungarian Empire, and Germany during the two world wars, followed by famine and Chernobyl. The City of L'viv, known variously as Levensburg, Lemburg, Leopolis, L'vov, and its people, has survived them all, and is positioned to thrive as Ukraine develops its transportation system and its businesses to attract more interaction with its neighbors in Europe.

For fifteen years, I have been telling our friends stories about this journey. For five years they all said stop talking about it, write the book. For the past five years I have been working on this book. This story is about the journey, and stories about the place, the people and the corporations or organizations for which they worked. It is about friendship and betrayal, weakness and strengths, success and failures, and a friend who was shot.

Like the Ancient Mariner, I am compelled to tell this story. This is my story. I am not a professional writer. The writing of the story might not be the same as living it, but I have tried to write about it the way we lived it. The way we lived it was like the radical twists and turns of a river. We went forward and backward and sideways.

This is a life story, and we all know that life is real, which means that good and bad things happened to us. It is up to each reader to decide what meanings can be derived from this story. As we often do not judge a book by its cover, we often do not judge a life by its ending.

Parts I through VI are the business story. Part VII is about Ukraine, L'viv, and the life we experienced there. The separation of the two stories became necessary to make it easier to follow the business story. However, the reader can start by reading Part VII to learn about the country and the times.

The Founders

Bill Schlageter and I were friends from the beginning, and then became partners in our business. Bill, an outstanding man, friend, and partner, had been the chief Financial Officer at Michigan Bell, when he was promoted to Vice President of Network at Ohio Bell. This was the largest organization in Ohio Bell in the 1970's and 1980's with 11,000 employees in his organization.

Bill, 6'2", trim, no fat, disciplined, an active runner, an active sportsman, and graceful, but not athletic. His Catholic training taught him a love of grammar and finance. Bill created all the financial plans, business plans, letters, and documents on his computer and sent them email. Bill loved to edit, even letters that had already been sent! He was disciplined, without being stiff. He never swore, and never lost his temper. In my day, you would say he was calm. Today, you would say he was cool.

Whether you met Bill and his wife, Linda, in high school, college, work, or the communities where you lived, or were relatives, it did not matter. You were friends for life. They had a great commitment to their friends. They traveled together, played together, ate together, there was always something going on. One example of this is the Christmas ornaments Linda and Bill made every year. Often they involved Bill's carving wooden figures and Linda painting the Santa Claus which they gave to friends for Christmas. Their ornaments are prized possessions. Bill and Linda were always truly unique and inspiring

Bill and I became friends shortly after he and Linda and their three sons arrived in Cleveland. Bill and I were both avid fly fishermen and did a lot of fishing together. Bill was an avid duck hunter, and active in Ducks Unlimited. He had an extensive collection of antique duck decoys, and made his own wood duck decoys. But we never went duck hunting together for one simple reason. There was no distance that was safe enough between a gun and me. I could not shoot a gun. I could not learn to shoot a gun in basic training. All I got was a black eye.

We started the business as friends and ended up as inseparable partners, joined at the hip as my wife Kathy described it. The four of us even took a week's trip on a working ore boat from Detroit, Michigan to Duluth, Minnesota.

How many hundreds of trips did we take together? We went to London, Frankfurt, Athens, Kiev, L'viv, Moscow, Prague, Los Angeles, Boston, New York, Washington DC and Atlanta. Often we were up for as much as forty hours at a time without sleep, food, or bathroom. There was never once an angry word between us.

If we were successful business partners, I think it was because we focused on our strengths, never our weakness or bad habits. We took delight in how we complemented each other. We always saw the whole as greater than the sum of the parts. I valued his discipline and efficiency, and he was always prepared with whatever document or financial plan we needed. He saw the value of beating the strategy horse to death, one more time. We were cutting everything out of whole cloth without a pattern. Teamwork was our guiding principle. There was also the strong desire to learn new things, meet new people, and solve problems that initially we could hardly understand. This project was driven by our enthusiasm, and the clear understanding that if we had a problem, we had to fix it ourselves.

Except for the times when one or the other of us was alone in Ukraine, Bill and I did everything together. We worked together in the office, attended every meeting together— meeting with investors, the Bank and the manufacturer (Hughes). We did this so many times that Bill could deliver my speech, and I could deliver his. Many times, I diverted Bill from his work on the financial plan to discuss some strategy or some problem. No matter how many extra hours he would then have to work at night, he never complained. As Chairman of the Boards of the two organizations, Hatwave (the western investors) and Ukrainian Wave (the company doing business in Ukraine), Bill earned the total respect and confidence of everyone. No one could run a better meeting. He earned everyone's respect by listening and following through.

Before joining Ohio Bell, I had been the first law clerk for Judge Anthony J. Celebrezze, U.S. 6th Circuit Court of Appeals, who had previously served both President John F. Kennedy and President Lynden Johnson as Secretary of Health, Education, and Welfare (HEW) before being appointed to the Court of Appeals by President Johnson. Before that I had been a law clerk to Federal District Court Judge Paul Jones in Cleveland Ohio. Judge Celebrezze was a great mentor for a young lawyer, and we developed a lasting friendship. I then spent 24 years in the Ohio Bell

Legal Department. During the last several years, I had been on loan to the Executive Vice President working on several strategic initiatives of the company. For the last and the 25th year, I was Director of New Business Development.

I had spent many enjoyable years serving on community boards, and was active on many state and local political campaigns, always behind the scene. Several times, I was a precinct committeeman, but then someone would run against me and I would lose. I played a lot of tennis and chess, loved reading and music. My one claim to fame was that I "played" in our home on my beginners Artley Flute, complete with dents placed by my children, the equivalent of "Mary Had a Little Lamb" for the Master of the flute, Jean Pierre Rampal, who was a guest in my home after his concert in Cleveland. My only previous publication is a full page article on 35 years of ocean fly-fishing, published in the Martha's Vineyard Gazette.

Part II: Ohio Bell and New Business Development

1992 The Beginning and First Visit To Volgograd and L'Viv

In the late 1980's we believed there would be further consolidations in the former Bell system, and Ohio Bell would need to develop new business. By 1992 our idea was to take our telecom expertise and export our abilities to build and operate telephone systems in countries lacking the telephone infrastructure to provide service to all their people. At this time, however, no local Bell operating company was permitted to do international business. This was also true for Ameritech, the holding company that owned Ohio Bell. Ameritech owned a separate international telephone company. The first of many mysteries now occurred. The President and Executive Vice President of Ohio Bell agreed we could try. The President of the Ameritech operating companies also agreed

I left the Legal Department, and suddenly, for the first time I became a businessman, not a lawyer, with the title of Director of Business Development, working with Bill. This was uncharted territory for me. At Ohio Bell I was referred to "as the butterfly chasing his shadow", but one advantage of my ubiquitous activities was that I knew many people in the company. Also, for the first time in my life, I was the head of an organization with two people reporting to me: Barbara Sheers, my secretary, and John Chandler my assistant. Running an office and doing all the paper work were things that were a piece of cake for Barbara and John. John and Barbara were calm and steady. They were both well organized and effective

We began looking for sites to implement our idea. Among the possibilities were Cleveland's sister cities, Volzhskiy and Volgograd, Russia. I knew some businessmen from Volgograd, so we decided that was the first place we would go. Before I left for Russia, I received a call from Walter Bazarko, who was referred by a friend of mine, Common Pleas Judge Burt Griffin, who was familiar with the sisters' cities program, and our intent. Walter and I did not know each other, but he wanted to see me about going to visit L'viv in Western Ukraine. I said no for two reasons. I had only been to Europe once in my life, and I was over my head in just going to Russia. Secondly, I had absolutely no idea where Ukraine was. He came to see me, and he was persuasive. After my visit to Russia, I took a train from Volgograd to L'viv and met Walter there.

Traveling with me on this first trip was Paul Karas, an engineer from Ohio Bell who was from the Czech Republic and spoke Russian. He was an experienced traveler, as I was not. At first I did not understand his concern that I was paying too much for some local craft or art item. I also did not understand that it was not necessary for him to translate every conversation for me, but that those conversations often yielded useful information to him, which he shared. I did not understand at that time that words were not the only means of communication. Attitude and actions were all the more important, because we did not have a common vocabulary of words for communication.

There are two things that stand out in my mind from my visit to Russia, when I traveled to Volgograd and Volzhskiy. Volgograd was totally post-war architecture. This was understandable. One of the fiercest and deadliest battles of the Second World War was the battle for Volgograd, then known as Stalingrad. It is said there was not a meter of land of the entire city that did not contain bullets, military hardware, or bodies.

There were four cities of strategic interest to the Germans in their Eastern campaign: Leningrad (now St. Petersburg), Kiev to protect the southern part of the eastern front, Moscow, the capital and center of the government, and Stalingrad (now Volgograd) the entry for oil. That is why Hitler ordered General Paulis to fight until the death of the last German soldier. That is why the City was totally destroyed.

The first most moving experience was a visit on the banks of the Volga River to the memorial for this famous battle of Stalingrad. The memorial is a domed building. One walks a spiral staircase around the dome ascending some several stories as one viewed the murals of scenes of the battle. The aerial and ground battles were magnificently displayed. At the top of the dome was the scene of the German General Paulis' surrender. After such a bitter and devastating battle, the scene depicts General Paulis surrendering and still holding his weapon as a courtesy extended by the Russian General. Professional military soldiers fought the war, directed by their governments, but they had respect for one another.

In Volgograd, the politicians were disciplined, stuck to business, detailed and cautious. I appreciated that approach, but I was concerned about the central role of Moscow in the regulations of a telephone system.

The second most moving experience was when I left Volgograd. I was there for one week. I met with government and business leaders. I saw the Volgograd Orchestra and met the conductor. He and the orchestra had actually played in Cleveland. When I left, my host had a farewell party for me. He put his arms around me and cried. It took me many years to understand why. He knew he would never see me again, because he correctly believed I would not return to Volgograd to start a business. Because of my host, I have as a remembrance, matted and framed coins and rubles from Soviet Russia. My host signed one of the bills. I had another friend there who gave me a Perestroika watch

Then Paul and I went to L'viv. I was totally shocked. Walter and a large entourage met us when we arrived after a long train ride from Volgograd. Walter had really done his homework. He arranged meetings for me with the head of the Oblast (one of the 24 provinces of Ukraine) who assigned his third deputy, V. Kalynyuk, to work with me. I met the Mayor and local businessmen. I was asked to sign protocols or letters of intent, in the hopes that I would come back.

Finally, I was impressed by the City itself. It was a beautiful old city, not built after World War II with new stark buildings and apartments, such as we saw in Russia or as we saw later, surrounding the old cities in the newly independent states, like the area around Prague or Vilnius. And I was struck by the obvious respect that the people had for Americans. We felt we could do business there. Of course, we all know what happens to people when they fall in love.

Walter Bazarko was the reason I went to L'viv. Walter had a Master's Degree in Physics, and a law degree. Walter was born in Cleveland, but his parents were Ukrainian. Walter loved Ukraine. He spoke the language and he sang the songs. He was an active member of his church and the Boy Scouts, one of the few voluntary associations that had survived in Ukraine. I never met a Ukrainian American that loved Ukraine more than Walter. He preferred to give his name as Volodymyr rather than the American "Walter." I never met anyone who knew 2000 years of Ukrainian history better than Walter. Finally I never met anyone who could tell more jokes.

Every trip Walter took to Ukraine he brought presents. He brought needed medical equipment, and an endless stream of things for people, institutions, hospitals, or Boy Scouts. All of the initial people we met in L'viv were Walter's friends. Walter had a background in mathematics and physics, and had worked at NASA, before he became a lawyer. Walter's legal career had nothing to do with business. It was not that he disliked business; it was just never his thing. Walter loved people, art, Ukraine, and, of course, his beautiful and very successful wife Lydia and his children. Walter was a people person. His parties were always fun, and he was attentive to everyone.

Walter was tall, not athletic, but moved effortlessly. He had a magnificent baritone voice. Over the years, I spent countless hours with Walter, going to restaurants, bars, meeting his friends and listening to them sing. Being with Walter was a unique experience and a pleasure.

In the early stages of developing the business in Ukraine, Walter was indispensable. He translated many documents into Ukrainian. In most of the meetings, Walter translated. This was an arduous job, especially when he had to convey reactions, feelings, and conflicting expressions of approval or disapproval. Often I would say, Walter tell them this, and Walter would say, "I already did." Then there were the toasts. They took place during lunch and at the end of the day's meetings. The toasts were a critical part of doing business. They gave the opportunity to say good things to each other, soften the edges of a debate, and establish respect and friendship with each other. It was an art form, and you were expected to do it well. I must confess that I enjoyed giving these toasts. Accurate translations were indispensable. The Ukrainian side knew and trusted Walter and sent messages to us through Walter.

After my first visit to L'viv, Paul and I took a train from L'viv to Prague, the place where Paul had grown up. The gauge and the width of the track in Russia and Ukraine are different from the gauge in Europe, obviously designed to make it difficult for troops to go from Europe to the Soviet Union. At the border, that night, the train had to stop. Paul and I stood outside the train and watched as they lifted up the entire train and changed the axels and the wheels.

1992 Ohio Bell Starts Business Development Group

As I have mentioned, I went to L'viv in early 1992, soon after Ukraine established its independence. One only needs to visit this city once to fall in love. What influenced me the most was the obvious fact that western businessmen would feel comfortable and at home doing business in the City of L'viv. Much of the city was Western style architecture. That is not to say that there are not many obstacles to overcome.

Because of Bill's on-going relationship with Northern Telecom, we talked with them about building a wire-based telephone company in L'viv. Over the next five months, I made at least four trips to L'viv. While I was there, one might say I communicated with Bill by telephone. It worked this way. Ukraine is seven time zones ahead of Cleveland, which means that three PM Cleveland is ten PM L'viv. I would call the local operator and ask for an international operator. Generally, "this was not possible." If I was lucky, within one to four hours an international operator would call and attempt to place the call. About one AM, still waiting for the call, I would give up and go to bed. This certainly made our idea about the need for our telephone system look good

The idea that "Ma Bell" would engage in a crazy idea like setting up an overseas company in a former Russian satellite, galvanized Ohio Bell employees. The enthusiasm and support swept through the company. Everyone wanted to help the project and did help. I have a red t-shirt given to me which says "Cleveland-Volgograd—Ohio Bell—Ameritech." I mentioned that Paul Karas accompanied me on my first visit, but there were others: Roman, a Ukrainian from our marketing department, and Richard Brown, my financial wizard, both of whom traveled with me.

But then we encountered a few obstacles. The first obstacle was that the L'viv Oblast, the equivalent of a county government in Ukraine, picked an engineer as their telecom consultant. He had a rare disease unlike any I had ever seen. Not even for a nanosecond could he ever look anyone in the eye. He was a young man, but thoroughly schooled in the old forms of communism. No one could reach an agreement with him on anything.

The business plan was to build a wire line telephone company in L'viv, Ukraine, a city of 800,000 with limited, antiquated local phone service. Ohio Bell would partner with Northern Telecom, a Canadian manufacturer of telecommunications equipment, which had provided equipment to Ohio Bell. Bill Cashell, a Vice President, assigned to cover the United States, had become a friend of Bill Schlageters. Bill Cashell agreed to send a team of engineers with me to determine if we could build a telephone system in L'viv at a reasonable cost and reasonable price for the consumers in L'viv. The team was terrific, and we thought we were making some progress. But then an Executive of Northern Telecom based in London and responsible for European Development came to L'viv to see what we were doing. The Executive flew to Kiev, and was driven to the train for L'viv by my friend, Nick Sergienko in his first car, which was brand new. The Executive managed to burn a cigarette hole in the seat, without apology or concern. I met the Executive at the train station with a delegation of Ukrainians and the Northern Telecom team. When he got off the train, after his twelve hour ride, he could not contain his anger. The indignity of having to ride the old train for twelve hours was only exceeded by the indignity of being required to come to a place like this, a City in the middle of nowhere. Within two days he left, and worse, he ordered the entire Northern Telecom Team to leave. He did not see that his company could have been first in the market of a city of 800,000, in Western Ukraine with a population of 10.0 M in desperate need of new telephone infrastructure.

The third obstacle was even worse, and the project crashed.

The Senior Vice-President and President of the Ameritech operating companies, a friend and supporter of this project, resigned. There was no corporate support for our venture. Bill and I talked it over. Well, all is not lost. We will just take early retirement and do it ourselves. So in 1993, that is what we did.

1993 Retirement

When Bill and I retired in early 1993, we were both provided with offices outside the company, each in a different building. Lou Masterson provided Bill's office to him and since I was always in Bill's office, Lou kindly offered me another office at no expense. Starting a new business was a legitimate strategy for outplacement, and Lou and his staff were very helpful.

Through Walter, we inquired from the officials in L'viv if we, on our own, could come back to build a wireless telephone company. We were told that Chaplea was long gone, and that yes, we would be welcome back. Without the support of Ohio Bell, we decided that wireless technology would be more cost effective and appropriate for a telephone system in L'viv.

On Bill's desk was a green metal file box with 200 blank index cards. By the end, we had two full index boxes of contacts we had made. After a thorough analysis, we developed a list of pros and cons. The negative list included no capital, no investors, no technology, and no manufacturer of equipment, no company, no financial or legal assistance, no business or financial plans, and no current agreements in L'viv or Kiev to start this business. On the pro list was our belief in each other and our belief we could do this. We decided to go with the pro list.

During the summer and fall of 1993, Bill and I began working on the negative list. The first issue was to form our company. Through relationships that Bill and I had with the International Law Firm of Jones Day, they agreed to take us as a client in their start-up business group. We formed our company, Communications Technologies, in which Bill and I were the two partners. Then through Bill's relationship with a partner in the Michigan office of Peat Marwick, they agreed to help us in raising the capital. Then Bill and I began the process of putting together a business plan and financial plan. We did a lot of research. We developed a rudimentary plan with many significant questions unanswered. This was a start. We were optimistic that 1994 would be the year.

Part III: The Beginning of the Journey in Ukraine

1994 Capital and Technology

In January 1994, Bill made his first visit with Walter and me to L'viv. On January 19, 1994, we entered into letters of intent with L'viv Oblast supporting our business proposal to build a wireless telephone Company in L'viv. V. Kalynyuk, who was now the 2nd Deputy to the Oblast, assisted us in this. When I was there in 1992, Mr. V. Kalynyuk was the Third Deputy of the Oblast. When I returned in 1993, he was the Second Deputy. He worked closely and diligently with us during the entire process. He became our closest friend and companion. He was known, respected, and a friend of most of the political and business leaders of the Oblast. V. Kalynyuk arranged most of our meetings and assisted in the development of the necessary protocols or letters of intent and agreement.

On returning to Cleveland, we began our search for the capital, the technology and a manufacturer to provide the equipment to build our network. We went to California to meet with Qualcomm to see if we could use their technology, known as CDMA. Qualcomm was not interested. Some years later, Qualcomm's CDMA technology would be adopted by three Ukrainian Operators including CDMA Ukraine, owned by International

Telephone Company, IT. We went to Chicago to meet with Motorola. They told us they were not interested in Ukraine or us. Soon thereafter, Motorola invested significant time and money to obtain licenses to operate in Eastern Ukraine, but it did not become a competitor, because when they lost their license, they left the country. We also met with several small manufacturers doing business internationally. They were not interested either.

During the first half of 1994, we made progress in Ukraine on the operating and frequency licenses we needed. We signed a second Letter of Intent in June. We received support in Kiev from the Minister of Communications for Ukraine. This was quite an accomplishment, since no one in Ukraine knew that we had neither the capital nor the technology yet. But we had to have the support in Ukraine if we had any hope of obtaining the capital and a manufacturer. These were the two big mountains we saw ahead of us, but in fact there was a whole mountain range, as we were to learn shortly.

During the second half of 1994, two significant events occurred which enabled us to climb the first two mountains. At the recommendation of one small manufacturer, we met with Dr. Arunas Slekys, Vice President of Hughes Network Systems (HNS), headquartered in Germantown Maryland. Dr. Slekys was of Lithuanian background, and had a personal interest in Eastern Europe. He invited us to meet with him and discuss our project. It was an impressive day. He had a strikingly rugged handsome face with black hair. He was medium weight and height, but strong build. He showed us a map of the world on his wall with hundreds of red pins representing places where HNS had done business. He informed us that HNS was owned by Hughes Electronics, which was owned by General Motors. One might remember the day when that meant something. HNS had working systems in Prague, Jakarta, Malawi, and Russia. Arunas was a super salesman, and most importantly, he was interested in us and our project. He wanted it to work and offered to help. We left this first meeting excited and confident that we had found a quality manufacturer who was interested in our project.

Hughes offered a proprietary fixed wireless system, which used a technology that no other technology offered at that time, 10-12 voice conversations on a channel, instead of the industry standard of 3-4 conversations per voice channel. It was a fixed wireless system, which meant that each home or office had its own antenna and equipment to send and receive the signal, and the customer had a phone that could be used anywhere in that home or office. This would give us significant economic advantage, which we needed if we were going to successfully do business in Ukraine. As luck would have it, they had a system built and available for us. They gave us their specifications for their system, a preliminary cost estimate, which we could use in developing our financial plan. It required only three cell sites to support services in L'viv for 15,000 customers. Bill and I were not only impressed, we were excited.

The list of potential capital investors prepared by Peat Marwick resulted in no interest from any of them. They were concerned about the country risk and how long it would be before they received a full return on their investment. In the fall, a close friend of Bill's, Jim Noteman, a partner from the Michigan office, set up a meeting in their Washington Office with Dagnia Zeidlakis.

At this meeting, she proposed a meeting in her office two weeks later with Mark Tomlinson from the European Bank for Reconstruction and Development (EBRD) located in London. Mindful that Bill and I had already talked with several international banks in Washington DC which resulted in nothing, I said we should not get involved with the EBRD because we would never cut through the long bureaucratic process. Dagnia was tall, beautiful, and she was a people person, who also happened to be very smart, very loyal to Peat Marwick clients. As one might guess, she ignored my comment. Mindful of the fact that we had not one dollar of capital, she said fine, you will come back here to meet Mark in two weeks.

As already discussed, Bill and I made significant progress in preparation for any meetings with capital investors, including this meeting with the EBRD. We had the support of Jones Day, our law firm, and Peat Marwick, our accounting firm. We had the support of Hughes, the manufacturer of the equipment. We had business and financial plans. We had a protocol of support in Ukraine. We were prepared to meet with Mark.

In two weeks, Bill and I met Mark in a large conference room. At one end of a large conference table sat Mark and Dagnia. Towards the other end, sat Bill and me. For over an hour Mark sat silently listening to Bill and me talk about our plans. At the EBRD, Mark was in charge of several countries, including Ukraine. He knew the country from other dealings. He was brilliant, respected, and of course influential. He was young, tall, slender man. When we had finished, Mark simply said, I want you to immediately file an application with the Bank, and we will support you. With Mark there was clarity and decisiveness. When Mark made a commitment, you could take it to the bank. There was no showboating, zero bureaucratic language, zero pointing out the million questions inherent in a new idea, and zero effort to show how smart and important he was. Mark simply said, submit you proposal and I will support it. How simple and basic, and yet so rarely understood or followed. We were certainly fans of Mark and Dagnia's when we left. Several Months later, January 1995, we submitted our proposal to the EBRD.

The Technology

Central to this story is the wireless technology we used and the telephone system we deployed. During the 1980's and 1990's, there were basically two technologies to deliver telephone connection. The first of course was wire, a land line system. This was the historical backbone of the telephone system we have today, but today it is broadband with the use of fiber optics and wireless which greatly enhanced capacity for data transmission. Back then to build or expand or replace a wire system with fiber optics was expensive because every customer has to be physically connected to the system by wire. Issues like the density of population, terrain, and the initial cost of installation required significant long-term capital investment. The second drawback is that one is tied to the geographic location of the system.

The second technology was wireless. With this explosive growth of cell phone service, the capital cost of providing the service has declined significantly over the last several decades. With broad band capacity, it has grown to accommodate data transmission. Its growth in this country and around the world has been spectacular.

But back in the 1990's, there were very limited wireless technologies available. Nevertheless, in countries with developing economies, and limited or obsolete wire line infrastructure, a wireless phone system seemed to be the quickest way to expand telephone service. Within the wireless technology, there were and there are two basic technologies to sending the radio signal on which the cellular phone system is based, "Time Division Multiple Access" (TDMA) and "Code Division Multiple Access" (CDMA). In both technologies, the audio signal is digitized. At that time, there was debate over the merits of each technology, but at that time CDMA was new and the outcome of the debate was not clear. We had to make a decision.

In TDMA, the signal is divided into a number of milliseconds long packets and then distributed to open time slots available on several channels and then reassembled to create a coherent telephone conversation. This increased the capacity of each radio channel by 3-1, and with today's technology, 6-1. In every voice conversation there are pauses which create the opportunity to put other parts of a conversation in the open time slot, thus increasing the capacity of each voice channel. In Europe and Japan, the TDMA technology is called GSM. It is the same technology, but at a different frequency than the USA technology, used by ATT and other regional phone companies. In 1993, TDMA technology was the only option available to us.

The other technology CDMA uses an assigned or specific code for each digital packet which is sent with a unique key. This is a spread spectrum technology that allows multiple frequencies to be used simultaneously. This technology was developed by Qualcomm, and was new to the field.

But back in the 1990's there was still a third cheaper alternative called "fixed wireless." Some of the transmission and receiving technology was located in the customer's premise, thus avoiding the significant cost of "locating" multiple cellular telephones moving in a defined geographic area. The fixed wireless systems used radio technology, but to fixed locations, requiring fewer cell sites or towers. A single residence or site could be served with a single subscriber unit (SSU). In an office building or multiple residences (apartment house) the entire building could be served by a multi-subscriber unit (MSU), with significant cost savings. From the SSU or MSU the call or signal would go by radio to an intermediate cell tower which then would send the call or signal to a central tower on a building that housed the switching equipment to switch the call and then to reverse the process and send by radio to the equipment located on the customer's premise. There was also the potential to have limited portability or mobility confined to a limited area, like a building, residence, or immediate neighborhood. At the time, this seemed like a cost effective alternative for the development of service in a City like L'viv, with poor wire line infrastructure, and good market for the development of an inexpensive, functional fixed phone system.

In 1994, Hughes provided a fixed wireless system, which was based on TDMA technology. However, it used a proprietary system called Extended TDMA, which according to Hughes, increased the capacity of each voice channel from 3-6 calls per channel to 10-12 calls per voice channel, obviously a dramatic cost advantage for us. Since the Hughes system was proprietary, not in the public domain, that meant that only Hughes knew this technology and it could only be bought by buying the Hughes system. This made it difficult to check the reliability of their system. In 1996, Hughes had received recognition for their technology. Our financial plan was based on Hughes preliminary cost estimates to buy and install the system.

There were several reasons for us to buy their system. First, HNS had a system already built and ready to install. Secondly, they were the only company eager to do business with us. Time was critically important. If we could start within the next year or two, we would be the first competitor in Western Ukraine, a region of ten million people. Most importantly, we would have a significant cost advantage with their announced capabilities.

While we were developing our plans in the early 1990's, the public telephone system in Ukraine, Ukrtelecom, was making its plans. In 1991, Ukraine inherited an infrastructure for telephone service that was analogue, not digital, based on metal cable, and had a waiting list for telephone lines of 3.5 M. It was one part of the Ministry of Communications, which was also responsible for the postal system. Ukrtelecom was separated from the Ministry of Communications, and began in 1991 to build a data transmission service, Infocum with Deutsche Controlware. In 1992, Utel was founded by Ukrtelecom to develop and provide long distance and international capacity. It was a joint venture with ATT, Deutsche Telecom, and Dutch KPN. The Utel partners were bought out in 2005.

1995 Capital and Licenses
Bill and I expected 1995 to be a decisive year. Mark from the EBRD asked us to immediately file an application to

the EBRD for a $15.0 Million dollar loan. The Loan required $6.0 M of equity from the western shareholders, and the equivalent from the Ukrainian shareholders. The critical step was to obtain concept clearance from the EBRD. After concept clearance, it was just a march through all the details necessary for final documents. We were told that concept clearance could take up to one year, not a happy thought. Our company, Communications Technologies, filed the application in January 1995. Soon thereafter, Mark informed us that the application was so good that the bank gave us concept clearance. It was formally issued in June.

Immediately after filing our application, Mark assigned Carl Gage to our project. Three days after being hired by the Bank, Carl met Bill and me at the EBRD in London. This was my first visit to London. Carl was an American banker and politician who decided to broaden his experience by moving his family to London. Carl was a friendly man with a big smile. He was eager to make this project work.

During the summer, Bill and I were in Washington D.C. to attend a meeting at the US Department of Commerce. At the reception we received an award. It was not really an award; it was support. Two men came up to us and introduced themselves. One was the Chairman of the Board of the EBRD, Dellarosia, and the other was his chief of staff, Ron Freeman. The Chairman said, "I know about you and you project. I want you to succeed and if you have any problems with my Bank don't wait, call me." Bill and I were shocked and elated.

Bill and I had already developed the business plan. The difficult part was the financial plan which Bill was putting together. This was a difficult and time-consuming process, done entirely by Bill. Over the year, Carl and Bill talked about the financial plan almost every day. If Carl made 500 changes, he made 1000. Carl received a lot of "help" from other bankers. Carl was a banker, but he also understood business and also had common sense. The other bankers thought only as bankers. So they drove Carl bananas. One change in the financial plan could result in changes in other parts of the plan. Day and night, Bill's nose was constantly in his computer, making changes in the financial plan.

It did not take Carl long to discover another problem. We were on a track to obtain $15M of loan or debt money from the Bank. How much of the $6.0 M of capital or equity money required by the Bank had we obtained? The answer was none. That was not going to work.

Off we went. We started with a venture capital firm in Cleveland. The director of the fund told us if he took our proposal to his board, he would be fired. We then flew to Hollywood, California to meet with a venture capital fund. Their team spent an entire day with us. While they were interested, the answer was no. Then we flew to Boston. Same result. They were interested in the project, but were concerned about the Bank's demand that the equity investors were required to guarantee the Bank loan, thus doubling their risk. This would turn out to be a significant problem. We went to New York to meet with a Japanese company. Same result.

Back to Cleveland, and back to Carl. Carl decided he could solve our problem. He did it this way. The telephone company, OTE, is the largest government-owned business in Greece. It had already invested in other countries in the area, Armenia and Romania. For over a year they had two large proposals with the EBRD. Neither one of them had received concept clearance. Carl called OTE and told them there were two guys from Cleveland, Ohio who knew what they were doing, but did not have any money. Carl suggested they could help themselves by investing in our business. They agreed to invest $4.0 M. in the project.

Then Carl called Hughes, and said if they wanted the Bank's help in buying the Hughes equipment for this project, then they had to invest $1.0 M of equity in the business. They also agreed.

Then Carl called Bill. "I have good news and bad news. The good news is that OTE will invest $4.0 M, and Hughes will invest $1.0 M." "That's good news, Carl," said Bill. "What is the bad news?" "CT will have to invest $1.0 M." "We can't do that," said Bill. "Bill," said Carl, "I have reduced your $6.0 M problem to $1.0 M. Do it!" With unbelievable faith and loyalty of some of our friends, we raised $800,000.

While all this was going on, we still had another large mountain to climb in Ukraine. We did not have the license to operate in Ukraine. Back to Ukraine, we went. Between January and July, the Letter of Intent from 1994 was upgraded to a Protocol and letter from Mr. Prozyvalsky, Minister of Telecommunications for Ukraine supporting Communications Technologies. There was agreement that the State Property Fund, Ukraine Telephone Company (Ukrtelecom) and the Railroad of Western Ukraine would become the Ukrainian shareholders in the business, with in-kind contributions equivalent to $6.0 M. Davymuka, the former head of the L'viv Oblast, would represent the State Property Fund. Kirpa, the head of the railroad, would represent the railroad. Ukrtelecom was represented at this time by S. Dembicky.

V. Kalynyuk, the L'viv Oblast representative who had been involved in our planning from my very first visit to L'viv, introduced Kirpa, the head of the railroad, to me. He was clearly the major player. In May 1995, a Protocol was signed supporting the new joint venture. A letter from Mr. Prozyvalsky was again sent supporting the Protocol. Local and national government support had to be given to every step.

So what we came up with was a Charter fund or initial capital investment of $12.0 M. In addition to the Western Investment by Hughes, OTE and Communications Technologies, this included $6.0 of contribution from the Ukrainian Shareholders in the form of space for our headquarters offices in the railroad building and space at Ukrtelecom for the switch and computers, the core technology, and locations for our three major antennae or cell sites. It also included equity for local assistance provided in the early days to obtain the necessary licenses from a small Ukrainian privately owned company, Renaissance DR. It included $15.0 M. of EBRD debt, for a $27.0 M total investment to develop a fixed wireless telephone system with Hughes equipment, in the City of L'viv. The first phase was to serve 15,000 customers, and in subsequent phases to grow to 100,000 customers.

To further assist us, Jerry Grant from Peat Marwick and I traveled to Athens to negotiate with OTE concerning their participation and CT's ownership in the western company where all the capital or equity money from the Western Investors would be paid. OTE agreed that CT would receive 21% of the western company for putting the company together.

Also of great significance, Jones Day Cleveland transferred

the representation of CT to Karl Herold, the partner in charge of the office in Frankfort, Germany. I could never imagine how significant this would become.

1995 Expanding Role of Kalynyuk and Kirpa in Obtaining The Capital and Licenses

As we have seen, during the critical years of 1994 and 1995, we had two core problems. The capital investors, OTE and HNS and the EBRD believed we would deliver the necessary licenses to do telephone business in L'viv. The Ukrainians believed we would deliver the capital to finance the business. While we were working hard to solve both problem, and were making progress, in reality we had neither firmly in place. It was like placing one set of cards against another. It was a delicate balancing act.

The hardest and most critical problem to solve was obtaining the operating license and the frequency license, which needed to include 10.5 MHz spectrum. These licenses were issued by the government in Kiev. We were unable and would not, just pay money, because that was illegal. For these reasons and for good business reasons, we decided on a bottom up approach. We would obtain the support of the people where we intended to do business. The local support would assist in obtaining the license in Kiev, the Capital of Ukraine. The first deputy of the L'viv Oblast, V. Kalynyuk, worked with us. He helped us in L'viv. When the company started, he was one of the two Managing Directors, along with Jerry Merkelo. He brought into the group H. Kirpa, who became a member of the Board of Directors.

V. Kalynyuk was a short stocky man, and like me and

Arunas Slekys, had a mustache. He was clearly a people person. He spoke little English, but his English was much improved over the years. In Ukraine, rarely would anyone entertain a guest in their home. From the beginning, V. Kalynyuk and his lovely wife Dozia, and her mother entertained Walter, Bill, me, and my wife Kathy, always in their home. No matter how many years they did this, we were never prepared for one challenge; the third course or the fifth course was clearly the main dish of the meal. It never was. When the main dish arrived, one wondered how it could be humanly possible to eat one more main dish. Fortunately, we managed.

V. Kalynyuk and I shared a common love of music. V. Kalynyuk established a tradition that lasted for every visit that I, Walter, and eventually, Bill, and Kathy spent in L'viv, attendance at the Opera. It is no surprise that the manager of the L'viv Opera House was a close friend of V. Kalynyuk's. The L'viv Opera Company consisted of 450 resident artists, including the orchestra, ballet, and the opera. The Opera House is one of the finest in Europe, built during the time of the Austro-Hungarian Empire. Directly above the stage was a box suite used by the Emperor of the Austro-Hungarian Empire. It was here we were seated. We entered the box seats from a large paneled room. It was a unique and special privilege. The Emperor's box was always available for our use. The view of the hall, the works of art, the statuary, the gold leaf and brilliant colors were breathtaking. Across the front of the hall was a large ballroom called the Hall of Mirrors, with several beautiful crystal chandeliers.

The choreography of the ballets were simple, sometimes austere, classical Russian ballet, and very acrobatic. The ballets and the operas were always professional performances, often spectacular. The best performance we have ever seen of Madame Butterfly was in L'viv. The Opera did mostly classical opera and ballet. Mozart was a favorite, but there were occasionally performances of Ukrainian opera or ballet. I must have gone 50 times. On one occasion, I was invited to a birthday party for the Managing Director. Fortunately, I was prepared. I gave him a colored engraving by BAK of an original costume he designed for the Bolshoi Dance Company. I also gave one to V. Kalynyuk.

Over the years, I took many trips with V. Kalynyuk and his family. One weekend he asked if I would like to go to a wedding of the daughter of his friend. After the church service in a small rural orthodox church we went to the reception at the family farm. There was roast pig, lamb, and enormous amount of food and drink. Of course, I danced with many of the ladies. I was asked to give a toast. In addition to thanking them for the privilege of being invited, I said, I hope with humor, that I brought greetings from the American People and the President of the United States. I'm not sure they knew what to make of that, but they were pleased.

On another weekend we went to a camp in Ivano-Frankovsk for a family reunion. We visited the family cemetery. People were constantly taking me to see historic sites, castles and historic villages. Twice I traveled, once by car and once by train through Western Ukraine, south along the Carpathian Mountains to Muchacovo and Uzhorod, which is five hours from Budapest. The famous black soil, the lush green fields, people working n the fields or washing their clothes in the streams, were as beautiful as one could imagine. People worked the field by hand or hoe. I never saw a tractor. This was the Achilles heel of Ukraine. There was limited or no refrigeration and a severely limited transportation system to get the goods to market. They took their goods to the nearest railroad or town.

On one occasion, V. Kalynyuk gave to me, Bill, and Walter, two red books. One was the application for membership in the Communist Party and the other was your membership book, an interesting historical memento.

On another occasion, V. Kalynyuk took me to a health spa with mineral water. By now, it was in decay. A soldier told me the following story. The Ukrainians took Premier Brezhnev to this spa. Outside in the field was a frog pond. The soldier told me Brezhnev's daughter complained about the croaking of the frogs keeping her awake. The noise problem was solved by instructing the soldiers to shoot every living animal in the pond.

During 1993-1995, V. Kalynyuk worked diligently for us in L'viv. He also introduced to us, and brought into our group, H. Kirpa, the head of the Railroad for Western Ukraine.

Kirpa was a heavy hitter. He was a man of medium build, black wavy hair, and very focused. He was a commanding person for two reasons: as head of the railroad for Western Ukraine, he had his own empire, transportation, money, hospital, and lots of employees. The second reason was that he was smart, arrogant and dominated a room. He was also very observant and an excellent listener. He commanded respect and was feared. His portion of the railroad system was the lucrative route from Poland to the north and Hungary, Czech Republic, and Slovakia from the West through the Carpathian Mountains. The railroad operated day care centers and hospitals for their employees.

Our first meetings were held outside of town in his motel complex. But he soon transferred the meetings to his headquarters building and to a large conference room next to his office. In the early part of the year, all the parties stayed in Kirpa's motel located outside L'viv in the country. We all would take a sauna after the meetings and dinner. In one of the early meetings there, Dagnia from Peat Marwick, attended. Before we left, I instructed everyone to bring their bathing trunks. So there we all were taking our sauna with Dagnia and wearing our bathing trunks. Later in the year we switched to Kirpa's conference room in the Railroad Headquarters building.

To demonstrate the value of the Hughes system, Hughes invited Bill, Walter, me, Dagnia Zeidlakis, and Carl Gage, the banker, to visit one of their systems installed in Russia, near Volgograd. Since Carl Gage was going to L'viv with us for an important meeting, we decided to combine the two meetings.

We all met at the International airport north of Moscow. We were met there by Arunas Slekys, V.P. Hughes. There were two cars waiting to take us to the domestic airport, south of Moscow. We had a little more than an hour to make our connection. Carl Gage was invited by the Hughes group to ride in their car. The four of us went in the second car. Our driver dropped us off at the airport and left. Our first challenge was to find where we checked in, a significant challenge. Finally, in the last building, we found a counter that offered the only possibility to present our tickets. There was a heavy set woman behind the counter who looked at our tickets. She then spoke, perhaps the only

English word she knew, "Impossible." I pointed to a plane outside the window. Angrily, she said "Impossible" and then left. We never saw Arunas, Carl, or the Hughes team. They were on the plane. We walked out of the building to a totally empty parking lot. Standing in the parking lot we laughed. We had no place to stay, no rubles, no one to take us anywhere, and no visit to see the Hughes operating system. Since we were going to Ukraine in two days, joined by Carl, we had to wait for Carl's return.

Out of nowhere appeared a tall man who asked us if he could help. We took the only safe choice, the National Hotel in Moscow, near Red Square, too elegant and too expensive. The next day we walked around Red Square taking pictures. I have a beautiful picture of Dagnia standing in the center of the Square. I am still puzzled why and how this happened.

Our investors left it to us to negotiate the Ukrainian Wave Shareholder's Agreement. The four of us and Carl were going to L'viv to accomplish this. The critical meeting to negotiate the Agreement was to take place in Kirpa's conference room. The conference room had a long conference table with at least fifty chairs. All the future Ukrainian partners were there. Representing Hatwave was Bill, Walter, myself, and Dagnia. Representing the EBRD was Carl Gage.

The Ukrainian Shareholders included the State Property Fund, the Railroad, and Ukrtelecom, the public phone company. There was a small private Ukrainian company owned by O. Kryskiv. Bill and I brought all the Ukraine shareholders together in L'viv to negotiate the joint venture agreement in July 1995. We invited Carl Gage because of the importance of the meeting. When the four of us arrived by plane in Kiev from Moscow to take the train to L'viv for the meeting, a government official met us. No going through customs for us. This was the only time that happened. We were taken directly to cars, and along with our luggage, driven to the Railroad station. We were taken directly to Kirpa's private car attached to our train. When we entered the car, Kirpa greeted us all. We were also introduced to his personal chef. The first thing we did was have a little drink of vodka. We finally had an excellent dinner, but I have no remembrance of what we ate. Bill and I missed the most exciting part of the evening. This was so because Bill wisely went to bed about 10-11 P.M. I missed it because I fell asleep. Sometime after that, Kirpa got up walked across the top of the table to pour some vodka for Carl, "The Banker."

The entire next day was spent in Kirpa's conference room discussing the many issues raised by the joint venture agreement. By mid-afternoon, we had come to the critical issue. At that time, Ukrainian law required the Ukrainian side to have the majority shares in the company. Renaissance DR was a private Ukrainian company that owned 4 shares, enough of the company to provide 51% Ukrainian ownership, but not 51% public ownership. This was a solution to the EBRD's requirements. Both sides had an equal number of seats on the Board, but the Western investors had put in all the cash, so the agreement proposed that if there was a tie vote, the Western side had the right to break the tie. All the cash was coming from the Western side.

The Ukrainian side vigorously objected to this, even though Carl Gage supported the necessity for this position. The debate became heated. Bill stood up and began to say that he would try to take this issue back to the Western investors. I pulled Bill's sleeve and said we cannot do this. It is not negotiable. Suddenly, Kirpa's hand came sharply down on the table. The clap shocked the room to silence. While he did not know what I said to Bill, he understood perfectly. Then he said to his partners. "Sit down. If Myron says it's not possible, then it's over."

Then Kirpa and I left the meeting and went back to his office. His office included a large desk and large conference table with every chair facing his desk. Behind his office was a small private room with a china cabinet. Kirpa does not speak a word of English. I could not speak a word of Ukrainian. For the next hour, while the meeting continued, we communicated with sign language and drank a little vodka. He demonstrated for me what he had learned in the military. At the bend of his elbow, he placed a full vodka glass. Without dropping the glass or spilling a drop of vodka, he drank the vodka. Although scheduled, we did not attend the opera that evening.

There were too many meeting, too many lunches, too many toasts to remember. But I do remember one lunch when I made a big mistake. The table was always u-shaped, with Kirpa generally at the center. Always on the table were many vodka and cognac bottles, but it was a breach of etiquette, in fact never contemplated, that anyone would refill his own glass. Kirpa's waiter would do that. In reality, there was never a need to do this even though there would be an afternoon meeting. There was never an empty glass. Now here is my blunder. One lunch, I observed that my vodka bottle was much finer vodka than Kirpa's. I got up and exchanged my vodka glass with Kirpa's. I noticed he was not pleased with this gesture. As soon as I drank his glass, I understood why. His contained water! I learned later that for health reasons, his son made a large bet with his father whether he would stop drinking and he did. But he wisely maintained the ritual appearance of toasts and drinking together, with water.

Early Life Lines

During 1994 and 1995, Bill and I were overwhelmed with work. We were negotiating Shareholder's Agreements with HNS, OTE, and the Hat Wave and Ukrainian Wave Shareholder's Agreements. We were negotiating the loan documents with the EBRD, and our employment contracts. Fortunately, life lines appeared.

The first life line was Dick Brown. He had retired from Ohio Bell, and was a good friend of ours. Dick agreed to work with the understanding he would get paid when the money came in. Dick had many of the same strengths as Bill. He was focused, methodical, considerate, truly an exceptionally fine person and loyal. Dick kept a to-do list, and kept driving us to complete items on the list.

The second life line was Karl Herold. At a breakfast meeting in Cleveland, Karl met me for the first time. Karl wanted to know why we were not using him, given the significant legal documents which were being negotiated. As a lawyer, I totally agreed with him. But the answer was simple; we just did not have the money to pay him at this time. Karl said don't worry about the money, it would come, and working with the EBRD would be a new and useful experience for him. We agreed that we desperately needed him and he became actively involved.

During my career, I knew and worked with many attorneys.

Karl is a giant, one of the best. He is a big man, an athlete, and sportsman. His intellect, his loyalty, and his legal skills exceeded his size. Karl threw his abilities and his heart into our project. He ended up doing most of the legal work on every agreement in this project. He attended meetings in London, US, and Ukraine. I was privileged to become a friend of Karl and his family. Whenever I was in Frankfurt, I stayed in Karl's home. Karl has a way with words, and thus his comments on this book have become part of the Preface.

Karl hired Yuri Zaichuck, originally from Kiev, to work in his Frankfurt office. Yuri received two law degrees, from Kiev and from Notre Dame College in the US. Yuri was a true intellectual. He was fluent in English. Yuri was skilled in translating the English documents into Ukrainian. This was not easy because the Ukrainian language had not been used during Soviet times, and had not developed the sophisticated vocabulary common in English law. In fact, the legal system was based on Roman law. But English governed all the documents. Yuri had little patience for "stupidity." Yuri assisted Karl in drafting the documents, especially the Ukrainian Wave Shareholder's Agreement. Yuri was especially knowledgeable about the Country, the technology, and the law. Yuri became a close friend of the Stoll family, and visited us many times. Yuri's brother, Oleg, practiced law in Kiev, and gave us great assistance over the years. Oleg and his family became close friends.

The third life line was my wife, Kathleen. She saw two friends, Bill and I drowning in work. She had her own consulting practice. She volunteered to help us full time. While she saw how much time we were spending on the business, I don't think she realized it would become

24/7 for her. There was one humorous example of this. Bill and I had spent ten hours in the office. When I got home she heard me on the telephone. "Who are you talking with", she asked. "Bill." "How is it after ten hours with him you could possibly have anything left to say"? The investors were skeptical of the hiring of a wife, but she did the necessary things we did not have the time to do. She quickly earned their respect, and when they wanted an answer to a question, they would call Kathy.

Kathy attended all the meetings we attended and always went to all the Board meetings, and was in L'viv whenever I was there. She became a mentor to the UW employees, and with the only women on the UW Board appointed by Kiev. Later she was appointed to the Hatwave Board.

The last life line was Jerry Merkelo. He had been employed by Hewlett Packard, and had done international business for them. He was born in Ukraine, and spoke five languages. It was our plan and he did become the Managing Director of UW. Jerry attended many of our meetings and gave us invaluable advice. Jerry hired all our employees. He did an outstanding job of finding the best and the brightest. Most were young and out of college. The only exception to this was he hired Dachko as head of Security. On the job training would not work for this position.

How Communications Technology Operated

Central to this story, I believe, is we worked together, which was totally consistent from beginning to end. Our team began with Bill and me and extended to include Walter Bazarko, Dick Brown, and Kathy Stoll. One constant never changed the overwhelming number of problems to solve.

Every issue was discussed as a group. Everyone was treated as an equal. There was no adherence to the ancient principal of "SSK" (Source of Superior Knowledge and thus the source of power over others.) Our approach had always led to thoughtful analysis, consensus, and trust. Another critical consequence of these constant meetings about complex issues, we all developed a greater understanding of the complexities as they developed. For example, a suggestion to fix a problem in the financial plan could be answered by "this will fix that problem, but unfortunately it will create a worse problem elsewhere in the financial plan."

Another example, when Walter Bazarko introduced us to V. Kalynyuk and Oles Kryskiv in L'viv, and then suggested they become part of our group, there was no need of a discussion, they became part of our group. The best example of all them was over the Western Company. Hatwave was formed and the operating company doing business in L'viv, Ukrainian Wave, was founded, there was not a single discussion of who would be Chairman of the Board of both companies: Bill Schlageter would be Chairman and Myron Stoll would be Executive Vice President. We operated the same way, whether it was with the investors, the shareholders, or our employees. This approach made all the difference, especially when things became difficult. This approach is especially important for a small start-up company. We did not have the benefits of an outside independent Board of Directors. However desirable this would have been, we did not have the luxury of paying for the time consuming efforts of Board members. We had to depend on the team.

1995 The Investors

Once we completed the negotiations with the Ukrainian shareholders, we now had to negotiate with our potential shareholders, OTE, HNS, and our company, CT, to form the Western partnership that would establish the Joint Venture in Ukraine. At the request of Carl, Bill and I were invited to Athens. OTE arranged for us to stay in a magnificent hotel in downtown Athens, next to the parliament, and overlooking the Acropolis in the distance. The drive north from town to OTE's headquarters was crowded. There we met Vasilios Maglaras, head of the group reporting to the Chairman of OTE International. It was easy for us to understand that he had one time been a diplomat in Europe. The negotiating team included Manolis Georgakakis, Yanis Kaligirou, Costas Petrides, and two consultants, George Politakis and Dmitri Cocalis, an attorney. Manolis, or Manoli, as Arunas later called him, was head of this working group. This was the team that would stay together for the next several years. It took us several trips to Athens before they decided to participate in the project. We thought we had gained the interest and commitment of both OTE and HNS in the development of the phone system in L'viv.

There was one humorous cultural difference with the Greeks. They would always take us to restaurants that served outstanding fish dinners. They would ask us what time we would like to go out, and before we could answer, they would suggest midnight, not quite what we were used to. We compromised at 11:00. Of course at 11:00 pm the restaurant would be empty. Only when we left, the restaurant would be full. The food was always outstanding.

Over the years, there were changes in the OTE staff assigned to our project and we had many meetings with OTE in Athens. One pleasurable visit took place in 1996 as we completed the negotiations on all the corporate documents. Our farewell dinner took place in a restaurant directly across from the Acropolis. We could see as we looked out the window, on top of the hill, shining brightly against the night sky was the Acropolis. Although we had visited the Acropolis many times, nothing could compare with this breath-taking view. We were each given a bottle of red wine. As I write this, I am holding my unopened bottle. The label reads "Villanyi Cabernet Sauvignon 1996, bottled for the felicity of Mr. and Mrs. Myron Stoll—to your good health." Several years later in Athens, my wife was elected by the Western investors to the Board of the Western Investors' Corporation, now named Hatwave. They had originally strongly opposed her being involved in the business at all, because she was just a "wife."

It was a long difficult balancing act to persuade the Ukrainian Government we could obtain all the necessary capital to fund the business while at the same time to persuade the investors we would obtain the operating and frequency licenses necessary to start the business. These two efforts were rewarded in the last half of 1995 and the first half of 1996 when we received all that we negotiated for in the licenses and the agreements to start the business.

Part IV: Launch of the Business

1996 The Party

In Cleveland, Ohio on March 1996, we officially launched our newly formed wireless telephone operating company in L'viv, a city of 800,000 in Western Ukraine. In a conference room of the law firm Jones Day, overlooking Lake Erie, all the parties signed the Term Sheet for the European Bank for Development and Reconstruction, which established the basic terms for the ultimate Loan, the term sheets for the Shareholder's Agreements prepared by Jones Day, and the Master Supply Contract with Hughes. Representing the Ukrainian side from L'viv Ukraine were V. Kalynyuk, H. Kirpa, and S. Dembicky. Representing the Western shareholders were M. Georgakakis from the Greek Hellenic Telephone company (OTE], A. Slekys, representing Hughes Network services, the manufacturer of the system, Walter Bazarko the Ukrainian American who introduced us to Ukraine, and the two guys from Cleveland. In addition there were lawyers and consultants.

After signing all the documents, we moved to the foyer overlooking Lake Erie at Jones Day, and gave toasts. Toasts always involved a small glass of vodka, a lot of good words, and then bottoms up—the glass was emptied, and filled for the next toast. One might call it "vodka tasting, because the amounts were small, straight up, but finished at the end of each toast. That was the ritual. The Ukrainians always sang songs, and they did on this occasion. Now there were the unofficial celebrations, with three days of shopping, sightseeing, shopping, dinner parties, shopping, singing, toasting, and drinking vodka toasts.

Left to right: Walter Bazarko, Bill Schlageter, Stepan Dembicky, Manolis Georgakakis, Hryhoriy Kirpa, Arunas Slekys, Valerij Kalynyuk, Myron Stoll. At Jones Day law firm in Cleveland after signing all the initial documents.

The Ukrainians had been met in New York by Walter and driven to Cleveland by way of Niagara Falls. When they stopped at McDonalds to eat, and discovered they could not buy any alcoholic beverage, the guests drank their Hyrlka (vodka) in paper cups from a bottle in a paper bag. It probably was the first time vodka was served at McDonalds, though not to MacDonald's knowledge. Hyrlka was one of the first Ukrainian words learned by my wife, and the mere mention of the word would cause an immediate smile, at the common joke, which she used to advantage, although she did not like vodka.

There were three dinner parties in Cleveland Ohio to celebrate. The first evening we took the group to one of Cleveland's finest fish restaurants, overlooking the Cuyahoga River. We always had outstanding fish dinners when we visited OTE in Athens. On my recommendation, the group ordered Lake Erie Walleye. There are no worse alternatives than bad fish. To my horror, the Walleye was not edible. However the evening was joyous because the Ukrainians stood up in the restaurant, sang songs, and drank hyrlka. This was an uncommon scene in an American restaurant and caused some uncommon attention from our fellow quests.

Kathy took the group to visit NASA in Cleveland, where they enjoyed learning about NASA, and tested a demonstration of satellite telephony. Dembicky, the head of Ukrtelecom in L'viv was delighted!

The next day, with the assistance of Kathy and Walter, the guests from Ukraine and Greece took their lists and went shopping. That evening we ate a superb dinner at Bill and Linda's Schlageter's home, overlooking Lake Erie and downtown Cleveland. There is no finer hostess, no one more artistically creative, than Bill's beautiful wife, Linda. Their home with its view of Lake Erie is spectacular. It was no surprise; we sang songs, Ukrainian, Greek, and American. Kirpa played an accordion, and then Arunas Slekys played the accordion with one hand, and the piano with the other. It was a fun, family style evening.

On the final day there was more shopping. Our guests were now ready for the next big party. There were thirty six guests in our home for dinner. Included were some of our local investors and their wives. There were partners from our accounting and law firms. Jerry Merkelo was there as our first Managing Director. He had been a Ukrainian ex-Pat working for Hewlett Packard before his employment with Ukrainian Wave. We invited a local friend from the significant Ukrainian Museum in Cleveland, Andy Fedesky. We invited another friend, Maria Kaiser, who is an excellent photographer who took pictures of this event, the start of a brand new company. Several of our financial investors and their wives attended. When I attempted to give my toast of welcome in Ukrainian, it was translated into Ukrainian. There were toasts all around.

Kathy had asked the Ukrainians if there was anything special they wanted to eat. Kirpa asked for lobster. So in addition to an American meal of prime rib, potatoes, vegetables and salad, Kirpa had a 5-pound lobster. That created a near disaster during dinner. The stove, a six burner, two oven gas stove, was so hot that my daughter and her friends who were serving could not turn off the fire on one of the burners. Without thinking or hesitation, Kathy solved the problem by throwing cold water on the fire and the burner, and went back to her seat. Her neighbor, a gourmet cook and one of our accountants, Jerry Grant, volunteered the fact that serving 36 people at one time was a major culinary challenge, and before he finished his sentence, Kathy was back in the kitchen to turn off the gas!

At one table they talked about women's rights. At another they talked about a shipment of pharmaceuticals from Youngstown, Ohio that disappeared before it got to L'viv, a whole Railroad car never arrived. We had come together as a team. This truly was a celebration enjoyed by all. Now we were ready to work together, starting the business.

1996 The Launch

In April, after the meeting in Cleveland in March 1996 and the signing of the Term Sheet for the EBRD Loan, the term sheets for the Shareholder's Agreements prepared by Jones Day, and the Master Supply Contract with Hughes, the Boards of Directors of both Hughes and OTE approved their equity participation. In May, the EBRD's operations committee approved the loan. In June, Ukrainian Wave was registered in Ukraine as a Foreign Joint Venture.

By September, it seemed clear to Bill and me that we were about to scale the mountains. For the first time, we had all the kittens in the basket. The process, even though it took longer than we expected, we were ready to start the business. We had the financing, the licenses, and the equipment. It took two and a half years, but we still could become the first competitor in Western Ukraine.

Suddenly one kitten jumped out. In Ukraine, a new wireless telephone company, Kiev Star was formed. President Kuchma's daughter was the Vice President of this company. To assist Kiev Star, the President did what any father would do; he had all the existing wireless telephone licenses revoked. All the companies had to reapply for their operating and frequency licenses.

This meant new tests. For example, there were tests to demonstrate that our frequencies granted in the first licenses did not interfere with the military. We also had to pay expensive fees for the new licenses. Motorola, who had told us three years ago they had no interest in Ukraine, or us, also had a licenses and had begun building an operating system. Motorola did not receive a new frequency license, and left Ukraine, losing over twelve million dollars. In October, Ukrainian Wave was granted an operating license, but we still had to obtain a new frequency license.

While we were confident we were well on the road towards raising the capital for the business, it was not completed. To pay the costs incurred in obtaining new licenses and to keep us going Bill personally borrowed a substantial sum of money, to put into the company, and obtain the license.

In December, the frequency testing was completed and a positive report was issued. Also in December, the EBRD approved the UW Loan. We were still in good shape, we thought.

1997 The Investment

We received the approval of the Frequency Control Committee in March 1997 and finally the frequency license with 10.5 MHz of spectrum, the same as the first license. Ten months later, and having spent significantly more dollars, we were back where we were in 1996.

All the equity money came from OTE, Hughes, and Communications Technologies. The Western investors created the Hellenic American Telephone Company, named Hatwave. Hatwave would then become the western shareholder in the operating company, Ukrainian Wave. Since two of the three shareholders were American, OTE asked that Hatwave be incorporated in Cyprus. We agreed.

Since CT had done all the work in Ukraine, and was the

founder of the business, CT would be responsible for the hands-on oversight of Ukrainian Wave. We also agreed that Bill would be the Chairman of the Board of both Hatwave and Ukrainian Wave. CT negotiated employment contracts with OTE and Hughes for Bill, Walter, and me. But as part of the agreement, OTE and Hughes wanted the three of us to spend full time in Ukraine. Of course, we could not agree with this. This issue and our compensation became the subject of many months of discussions. While their proposal was not doable, the issue was legitimate, and a compromise was reached. One of the three of us would be in L'viv at all times, with the majority of the time split between Bill and me. So we rented a nice apartment. In fact, they were right. There is no substitute for boots on the ground.

In September, after these many delays, Hatwave agreed to the final revisions of the business and financial plans. In October, the Hatwave and Ukrainian Wave shareholders met and approved the EBRD Loan Agreement, the business and financial plans, the hiring of Jerry Merkelo, a former employee of Hewlett Packard with considerable foreign experience and of Ukrainian background, as the Co-Managing Director along with Kalynyuk. In November, Hatwave signed the Hughes Master Supply Contract. This allowed the equity investors to pay in their equity in December.

Despite the many surprises, despite the unexpected delays to get anything finally done, Bill and I replayed an old record: finally we would start the business. We felt good, relieved, and as always, confident. It was almost impossible to imagine or anticipate the shocks of 1998.

1998 Loss of The License, Local Politics

All my professional life, I had been involved in politics, both state and local. I knew how it worked. I was good at negotiating with the local and state governments. I knew nothing about national politics, however, and I had no clue on how Washington worked. If we were going to start a business in L'viv Ukraine, I reached one conclusion for several reasons. If we were going to do business in L'viv, I thought it was important that we go there and work with the local government. If the local government was on our side, then they could deal with Kiev. Why not start with the community where you intended to do business? The second reason was all other international companies had started in Kiev. There were several serious reasons why this was a bad approach, even though this was how many companies did it. First Kiev would tell the regions who they were going to do business with. Secondly, this approach required "paying" the right people." This we would not do, and never did. Thirdly, it was unique and appreciated by the L'viv government and the local businessmen.

During this time, I developed a friendship with an American businessman who stayed in Kiev. He had made his money in oil and gas in Russia. One day when he saw me, he said it broke his heart to see how hard we had worked, but we were losing out in Kiev because we did not work with the right people. He asked me to meet with a certain Minister. I said I would, but we were not going to change. I did meet with him and we did not change. Through the whole process, we met with the Ministers of government agencies, high level administrators and many government officials at the national and local levels. While we tried hard to make it difficult for them to stop our project, they could have stopped us at any time. They certainly made it difficult and time-consuming. But you could see it in their eyes, and how they dealt with us. Secretly they wanted us to succeed. They wanted to see if an honest business could succeed.

The strategy was a good one, and it almost worked. We had just started 1998 when the bomb fell. It was partially caused by the law of unintended results. In December, the EBRD agreed to finance the business, In December 1997, the equity was paid. Those two guys from Cleveland were going to make this happen.

In January, Kiev stopped us dead in our tracks. Despite all that had been accomplished and all the agreements signed, the threat was real and potentially fatal. This was a thunderbolt. We were again back to climbing that mountain. For the next two months, this became the only issue, and we had to solve it. Because the threat was so serious, we had only one chance to develop a sound strategy, and fire every gun possible. We had started by avoiding Kiev politics. Now we were in the center of it.

If extraordinary help does not generate passion, then the recipient is not deserving of it. The US ambassador to Ukraine was Mr. William Miller and later Ambassador Steven Pilfer, and the US Commercial attaché was Andy Bihun. I have never met a US Ambassador or a Commercial Attaché to any country. Ukraine was not an easy country to represent any foreign government or business. To take those jobs seriously is a difficult and heavy responsibility. They rolled up their sleeves and weighed in. To be blunt, they did it for two guys from Cleveland, small fries. They became directly involved. They did not pay lip service. They fought it to the ground. Andy Bihun was the lead. He participated and fought this to the end. He also brought Ukrainian people into the US Commerce office in Kiev who were extraordinary.

Because of the difficulties Ukraine had experienced with US Companies and the difficulties that US companies had experienced in Ukraine, wisely someone came up with the idea of the Kuchma-Gore Committee to monitor the activities of ten companies. Fortunately, and with a lot of effort, we managed to be one of those companies. This was important to us for two reasons. It helped protect our company and it helped to protect us personally. We were on the radar screen. We were less likely to have an "unfortunate" accident. It is incomprehensible to me, that with the change to a new administration, this program was dropped. Is it a strange thought to think before you add or drop a program the government would talk with the people affected by the program? I went to Washington, DC and discussed this issue with the State Department. I asked certain friends to talk with key people within the Ministry of Communications.

In March, Mr. Andy Bihun established a meeting with senior officials in Ukraine. I was pleasantly surprised to see Andy attend the meeting. That was really putting it on the line. They had to look him in the eye. Kathy had prepared a two-page list of all the licenses, and documents agreed to and signed in Ukraine.

The meeting was the most stunning I have ever attended. Ms. Phillipova, from the Office of the Minister of Telecommunications, was chair of the meeting. Her opening statement shocked me. She opened by saying "You have reached the highest level of our government. I want you to stop. We are here to listen to what you have to say, so please begin." Bill began with the two-page list prepared by Kathy in front of him. Quickly one of the government officials interrupted him and began to argue. Angrily, Ms. Phillopova stopped him. "I said we are here to listen to what they have to say. No one will interrupt them again." No one did. At the conclusion of Bill's remarks, there were no questions. Ms. Phillopova then said to us. "Please put your guns down. I promise you we will meet again in two weeks with our answer. We have listened and will consider what you have had to say." We left.

As promised, two weeks later we all met again. It was a short meeting. Ms. Phillopova said we want your project to proceed and succeed. We also would like to appoint two members to your Board who will be of help to you. The crisis was over. She did appoint two people to the Board, both outstanding and helpful: Ms. Gonchar from the State Property Fund and Mr. Orlenko from Ukrtelecom.

We were back on track. We had received zero help from our investors. We were again ready to get on with it and start the company. But 1998 was not going to be the year. Another bomb, equally unexpected and nearly fatal, went off.

In addition to the Term Sheet with the EBRD, the Loan Agreements and the corporate documents, there was the Loan Guarantee Agreement with OTE and Hughes to repay the EBRD Loan if there was a default. We were not directly involved with this because it was specific to OTE and Hughes. This agreement angered OTE for several good reasons. They put in four times more equity than Hughes. Hughes also was the only shareholder that immediately profited from its investment in the business from the sale of the GMH 2000 equipment, $5.0 M. OTE was paying the major portion of the loan guarantee, $4.0 M, but could not collect it until Hughes was completely paid. Hughes took the place of the Bank in the event of Default and could take over the management of the company. It did not look like a good deal to OTE. But OTE had signed the EBRD Term Sheet and the Loan Agreement. OTE had made a commitment to sign the Guarantee as written, but they had not yet signed it.

Bill and I were summoned to a meeting on June 30, 1998 in Athens. A new group at OTE had taken over the management of OTE International. Mr. Karaplis had taken over the finances for the company. He proceeded to tell us what a bad deal this was for OTE and that neither OTE nor its Board would approve signing the Loan Guarantee. There was little room for discussion.

Two bombs like this in one year were enough to rattle anyone's cage. There was only one thing we could do, ask for another meeting. This took place August 20, 1998 in Athens. This time, Dr. Slekys from Hughes and Carl Gage from the EBRD attended. All I remember was it was a difficult day. However, OTE signed the Loan Guarantee in September. They had too much at stake with the EBRD on other matters and could not afford to lose the EBRD's support on these other more significant matters.

It looked like a catastrophic year, but by December it was done. Every document was signed personally by the representative of the company and the Bank, in the presence of the Ukrainian Notary, stamped, and registered in Ukraine. I do not remember any elation or celebration. All I remember was shell shock and exhaustion.

Looking toward Rynok square and the Municipal Building in L'viv.

Left to right: Myron Stoll, H. Kirpa, Jud Kenney of HNS.

Kirpa's Railroad car, left to right: Dagnia Zeidlakis, Kirpa, Myron Stoll, V. Kalynyuk. Carl Gage, Walter Bazarko.

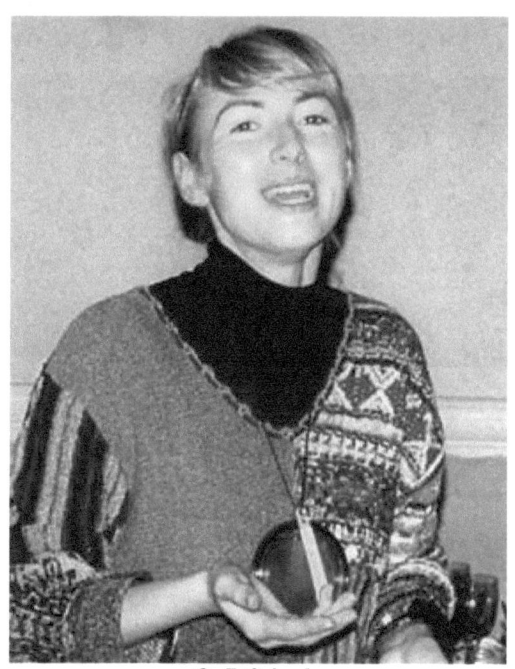

O. Bilichenko

L'viv Opera House, L'viv.

Party at Myron Stoll's house: left to right: Dr. Robert Blacklow, V, Kalynyuk, Winnie Blacklow, from right, Lydia Bazarko, Bill Schlageter.

Party at Walter Bazarko's House, left to right: Walter Bazarko, Bill Schlageter, H. Kirpa, and S. Dembicky.

Dagnia Seidlakis and Myron Stoll in Red Square in Moscow.

Plane with equipment in L'viv.

Party Myron Stoll's home, left to right: Myron Stoll, H. Kipa, S. Dembicky.

Linda Schlageter, Myron Stoll and Carl Gage.

Jim Noteman and Dick Brown

Part V: Starting the Business

1999 Shipping and Installing the Equipment and the Managing Directors

One would think by now that Bill and I had learned that this adventure was about climbing mountains in a rugged, uncharted mountain range. Except for experience and dogged determination, we did not anticipate what would happen in 1999.

The first problem appeared immediately in January. Since 1997, when Communications Technologies had purchased the Hughes equipment, it had been packed and held in a warehouse in big wooden boxes. On behalf of Hatwave, we were responsible for shipping the equipment, clearing customs, and paying the custom duties. In January, 1999 the EBRD disbursed two million dollars of the loan. Before the equipment could be shipped, UW was required to give customs a list of the equipment and its cost in order to determine the import duties. The estimates of the duties were ruinous. We asked Kathy to solve this problem. The first shipment included the switch manufactured by French Alcatel, the Hughes radio equipment, all the electronics, and cell site equipment. There were two components to all of this, both hardware and software. Hughes pricing bundled the two together. Kathy determined that the duty would be less if software and hardware were unbundled.

Since this system was proprietary to Hughes, they did not want to unbundle the software and hardware. Kathy, understanding the necessary solution, returned the favor of giving her this assignment by saying that since she had solved the problem, it was my turn to go to Hughes and solve it with them. It was a long, difficult day at Hughes. I was the one who had to constantly fight with Hughes. It was important for Bill to remain the good cop. I had only one argument; we had no choice, because the duty on the bundled software and hardware was ruinous to the business. Hughes finally agreed and unbundled the prices. Now a revised custom list was sent to customs.

This led to the second problem. Every crate of equipment had to be labeled and precisely match the custom list. It is interesting how many unexpected details can derail a project. Of course none of the crates were labeled and many were not properly packed, and contents had settled so the crates were not full. Fortunately, Bill and I had found an international shipper from Texas. He was a true entrepreneur, that is to say, he served the customers' needs in the safest, most cost-effective way. This resulted in the shipper, Kathy and Anatole, our translator, who together would unpack each crate, do an inventory, then label the crates and create a matching list in Ukrainian for each crate. They spent two days accomplishing this. The equipment was sent by plane to Belgium, and then to L'viv. Our shipper went the entire way with the equipment. On its way into L'viv, the plane was nearly hit by lightning. The tires were so worn, it was a miracle the plane did not crash on landing. No one was going to steal our equipment. Kathy, Anatole, and Dachko, the head of UW Security, and his staff met the shipment in L'viv. It was loaded on a truck for the trip from the airport to the installation sites, but it was not secured on the truck.

It was a miracle that it arrived safely after bumping over the hills and cobblestones of L'viv streets. It took eight men, all of them weight lifters, to carry one base station up eight flights of stairs. Little things were big challenges in Ukraine. The equipment arrived in March 1999.

The UW board meeting took place in L'viv March 24 and 25, shortly after the equipment arrived. Our next crisis arrived. Why wait and savor a milestone when one could have a real crisis immediately? During the March board meeting, both Merkelo and Kalynyuk, the two Managing Directors, gave presentations to the board. Something was going on below the surface. Both OTE and Hughes were angry with both Managing Directors. Bill, as Chairman, was directed by Hughes and OTE that UW was to have one Managing Director, who should be from Ukraine. They made it clear the new Director would not be either of the current directors, a major problem. V.Kalynyuk had been with us from the beginning and Merkelo had been with us almost from the beginning, and he had really gotten the operation started. He had hired staff, set up offices, and developed a budget. Merkelo had worked for Hewlett Packard, and had set up technical operations in other places. He had used local personnel and he knew what he was doing. Merkelo developed an excellent process to interview and hire the best candidates. Bill or I participated in many of the interviews of the candidates for employment. The employees, especially the engineers, and technical people that Merkelo hired were outstanding, and most spoke fluent English. Being the Managing Director of this company meant everything to Merkelo. But he lacked experience in reporting to a Board and they were not happy with his capital investments and expenditures which were going to create a cash problem for the company before operations could support the on-going operations. He had been accustomed to working for a large corporation with big pockets, which could afford to do things right. But he did not fully understand that the resources for this start-up business were limited, and had already been diminished by the year of delay in shipping and installation of the equipment.

I knew that firing V. Kalynyuk was not an option. It would create a firestorm in L'viv. The current and past mayor of L'viv, the current and past head of the Oblast, and the L'viv members of the Board, were all friends of V. Kalynyuk. And furthermore, to divide and split the Board was not a good way to start the company.

I had a new assignment. Solve it. The first thing Bill and I agreed was to take OTE and Hughes demand as a directive, not an immediate action. We told both Directors we were not going to do anything immediately. Within a month, the issue was forced to a showdown. V. Kalynyuk and I had been discussing this issue for a month. V. Kalynyuk set a meeting with me and his good friend, a UW Shareholder. A neutral translator assisted. We all knew the issue was complicated. Bill and I did not have the power to tell the Board what we would do or not do. While Hatwave had the right to break a tie, the Ukrainian's owned the majority shares. The only solution was a compromise. I suggested one. UW would keep V. Kalynyuk as a Managing Director at the same salary. But his duties would be limited to government relations, and community affairs. He would not be responsible for the operations or management of the company. This was acceptable, and both sides appreciated the effort to avoid a crisis. At least we now had one crisis instead of two.

Of course I still had a problem. We had not discussed a solution with Merkelo. Since I was in L'viv, Bill said I guess you draw the short straw. We worked for many years with Merkelo in establishing the company. He was a friend. He worked hard in starting the company, and did a lasting service by hiring outstanding employees. I knew we could finesse keeping V. Kalynyuk. Keeping Merkelo was not negotiable either with OTE and Hughes, or the Ukrainian side. Why the Ukrainian side did not like Merkelo was exactly why we appreciated and supported him. Under the Soviet system, the only protection people had was through family and friends. In hiring employees, Merkelo was under enormous pressure to do just that. When Merkelo proposed hiring a bright young engineer, I was told we could not do that. "Why" I asked? "Because we do not know him." Kathy came up with a clever idea. She wrote and the Board adopted a Mission Statement, Personnel Policy and a Procedure Manual and a Code of Conduct. This gave Merkelo the support he needed.

The issue had not been resolved by the time of the June meeting of the Board which took place in Prague. The ostensible reason was for the Board to see the HNS system in operation in Prague. The real reason was to provide the Ukrainians with the opportunity to have a free trip to Prague. This gave us a better opportunity to better know each other.

After the Board meeting, the law of unintended results again came into play. The Board instructed us to also hire a Director of Marketing. We hired a search firm in L'viv to find a good candidate for Managing Director. The owner of the firm was an American, and the former senior executive of a major company. He was a good friend of ours. He narrowed the candidates to two. Bill interviewed them first. Before Kathy and I left for Ukraine, Bill asked us to interview them. Bill said both were excellent, you choose. We interviewed both, and indeed both were exceptional. Suddenly, I saw a solution to two problems. I suggested we hire Ivan Griga as the Director of Marketing, and Olha Bilichenko as the Managing Director, replacing Merkelo. I called Bill and he agreed. We hired both. We had a solution.

The discussions with Merkelo were long and painful. Finally Merkelo said he would leave if he talked directly with Bill and heard the same things from him. I called Bill and said buy a ticket to L'viv. He did. After several long meetings, we agreed on a severance package. In August the Board met our new Managing Director of Operations, O. Bilichenko, and our new Director of Marketing, Ivan Griga. One strange thing, neither OTE nor Hughes asked why we still had two Managing Directors. Some things are best left unsaid.

One of the first official acts we expected from the Managing Directors was to sign all the agreements which were assigned by Hatwave to Ukrainian Wave, including our employment contracts. All the documents were signed by Hatwave and were transferred and accepted by Ukrainian Wave except our employment contracts. We had been working in Ukraine for several years under our employment contracts. O. Bilichenko had signed our contracts, but Kalynyuk had not. Both Managing Directors were required to sign contracts of this magnitude. Someone in a position of power had claimed he had been promised a significant sum of money. If he was not paid, our employment contracts were not to be signed. Neither Bill nor I had ever heard a word about this "promise." We would not have approved of any such payment if we had known about it, which we did not. We did not agree to make this payment. As a consequence, our employment contracts were not signed by both Managing Directors, so they could never go on Ukrainian Wave books. Consequently, we were never paid a cent for our work.

During the summer, until November the UW engineers were trained on the Hughes radio equipment and the Alcatel switch. Our engineers were very smart people, and learned how to use and repair all the equipment. They passed their initial exam for the training given by Hughes personnel at 100%. Their achievements were outstanding and a full credit to their selection by Merkelo. They were in fact frustrated by the unwillingness of the Hughes engineers to allow them to really understand and operate the system independently. They were constantly in communication with the central office in Germantown.

In September 1999, the remainder of the EBRD loan, $5 million was paid to UW. In October, the equipment was certified by the Odessa Institute. This was necessary before the company began operations.

During the fall of 1999, the UW Audit committee, which consisted of representatives of Hughes and OTE as well as Ukrainians, submitted its first audit report to the Board. While the audit report found all expenditures legitimate, the report also found that recapitalization of the company was critical. The financial plan had called for additional capital at this time, so this was not a surprise, but it did require reworking the financial plan.

In December 1999, Hughes put enormous pressure on UW to pay most of the balance owed to Hughes. Technically, under the Master Supply contract, they were entitled to be paid. We did not want to do this since the final certification had not been completed, and the financial condition of the company. Hughes insisted, and they were paid. By December 31, 1999, we had 16 customers. This created a cash crisis by February, 2000.

Part VI: The End Game

2000 Default

To understand what happened in the year 2000, it is critical to explain what happened in the first quarter of the year. On February 9th and 10th, the Board met in L'viv. The financial reports disclosed that although we had customers and some revenues coming in, that by the end of February, UW would be out of cash. After the payment to Hughes In December, this had been anticipated. It was consistent with the five year financial plan which called for additional equity investment and release of the second phase of the EBRD loan as approved to support the operations of the company and its acquisition of more telephone equipment to grow the business. We had the license. We had phone numbers. We had a functioning sales office. We had growing customers. And we had an operating company with competent personnel.

With the newly installed equipment, customers complained of cutting, clipping, and echo. Hughes staff said it was a problem with the Ukrainian language, the speech patterns, and the difficult hilly terrain. Final certification of the equipment by the Kiev Institute began in October 1999, but could not be completed until the system was in operation. The initial certification of the Hughes system had been performed by the Odessa Institute. The Kiev

Institute was responsible for the final certification of the installed system. The final report was issued in March 2000. The system was certified for a maximum capacity of serving 7,500 customers not 15,000 as incorporated into the financial plan, on which the equity investment and the EBRD Loan was based. Subsequent tests confirmed this finding. The certification process of the installed system proved to be honestly, thoroughly, and expertly done.

Serving the many apartment buildings by multisubscriber units was a significant cost-saving advantage over using single subscriber units. We could not use any of the multisubscriber units. These multi-subscriber units failed certification and were never shipped to Ukraine. This did serious damage to the marketing plan, the price per subscriber, and the financial plan. The fixed wireless system certified for only half the capacity of the contract with Hughes was ruinous to the business.

This news was compounded when Hughes sent notice that it was going out of the fixed wireless telephone business. While they maintained they would continue to support the operation of the company and what equipment we could use that they had in supply would be available, there would be no manufacturing of new equipment. It was the perfect storm.

O. Bilichenko and her staff increased sales to meet the cash requirements. The engineers worked diligently to meet customer complaints and improve the service. Management postponed payments on contracts in Ukraine, but managed to pay the staff, and keep the doors open.

And Bill and I pursued the commitments of the investors in the financial plan to give us the cash needed to keep the operations going, while we worked on a solution to the technology problem. But neither Hughes nor OTE had any interest in increasing or protecting their investment, and the EBRD was concerned about lending the additional money when it appeared that the UW might not be able meet its schedule for repayment of the first phase of the loan. None of them saw the opportunity to grow the company with the achievements of the UW operations, which had managed marketing and sales, managed the technical problems, managed with limited resources, established a reputation in the market, and had the assets of licenses, one hundred thousand telephone numbers reserved for future growth, and a top notch management team. Despite the setbacks, this was an impressive beginning for a start-up company. We now needed new money and new technology to make it work.

On May 23-25th, there was an emergency Ukrainian Wave Shareholder's meeting, called at the request of OTE. The Ukrainian shareholders requested that the meeting be held in Kiev. But the HNS representative convinced the OTE representative that they should not attend the meeting if the meeting took place in Kiev. It did take place in Kiev, and they did not attend and stayed in L'viv. We were unhappy with that decision, so Bill, Kathy, and I attended the meeting with the Ukrainian Shareholders in Kiev. The Ukrainian Shareholders agreed that if new capital or new equity investment were made in the company, that their equity would be diluted proportionately, but to no less than 26%. This was necessary for any new investor to consider investing in the company's development. We urged a plan to refinance the company, bring in new technology and equipment, and to continue to use the Hughes equipment by moving the Hughes system out of the city into the Oblast. The Hughes equipment was now obsolete in the expanding market of cellular phones and Hughes was out of the telephone business entirely. OTE had no interest in investing additional money to develop the business, and the EBRD had no interest is saving the company. New investors did not want to invest as long as Hughes and OTE still owned shares of the company.

So in August, UW defaulted on the loan, and, without discussion, OTE and Hughes, under the terms of the loan agreement paid the EBRD in full. By the terms of the loan guarantee, even though HUGHES paid less than one third of the amount, Hughes became the preferred creditor, stepped into the place of the EBRD in all the loan-related documents. They had the right to take over the company and to be paid for its equipment before OTE received a return on their loan. The loan documents contemplated that Hughes as the manufacture of the equipment would be the major creditor of the Company by supplying the equipment as the Company grew. It was not contemplated that Hughes would go out of the business and would be one of the causes for its default. In August, someone came from Hughes to UW and announced that Hughes was taking over the company, and that he was in charge. In case there were any misunderstanding, it was made clear that Bill and Myron were out. But Hughes itself was out of the telephone business, so in one week, the Hughes representative left and never returned.

In my possession is a memento, ironically given to me and Bill, dated October 3, 2000. It is a plaque which says "Hatwave Hellenic American Telecommunications Wave Ltd. Investor of the Year '99." "Strategic Priorities L'viv." Ukrainian Wave was the second largest company in L'viv at the time, second only to the local tobacco company.

2001 The End of The Journey

O. Bilichenko was introduced to the Board of UW in September 1999 as the Managing Director of Operations. She ran the company. O. Bilichenko was tall, attractive, and articulate in English and Ukrainian, very thorough in her work, extremely smart, creative, and worked tirelessly. Since she related well with people, she did not wear her intelligence on her sleeve, and she was most effective. When she started, the system was in the final stages of installation.

What none of us anticipated was by March 2000, the company would be insolvent. By August, UW had defaulted on the EBRD Loan, which was paid in full by the guarantors, Hughes and OTE. Hughes was out of the fixed wireless business, so no further help would come from them. Since this was a minor investment for OTE, OTE would not make any further investments of time or money in the company. Bill and I negotiated with OTE that O. Bilichenko could continue to run the company from internal funds but not add additional debt or encumbrances on the assets. That is what she did.

Fortunately, the creative Ukrainian engineers were able to expand the capacity of the system by buying used equipment from other Hughes customers who had abandoned the Hughes system. They were able to expand the services to 13,000 customers, in spite of increasing competition from mobile services of UMC and Kyiv Star, and wire line services like Farlep and Ukrtelecom itself.

This is how the company functioned. Except for two short technical Board meetings, one in 2001 and one in 2003, there would never again be a UW Shareholder's meeting.

Bill remained the Chairman of the Board. Bill, Kathy, and I remained in constant communication with O. Bilichenko with weekly emails, and regular strategy conference calls. We were never paid any of our salaries. Bill's personal loan was never paid. We, and all the investors, including our friends, lost everything.

We tried everything to avoid default to the EBRD. In fact, in the spring of 2000, in preparation for Tranche 2 of the EBRD loan, we had prepared the business case for the required additional investment from OTE, and demonstrated that the company could be successful. But the EBRD, OTE, and Hughes had no interest. OTE and Hughes informed the bank that they would default on the loan when it was due in August of 2000, and paid the loan guarantee as required. The EBRD was out. Hughes never took over, as they could have under the terms of the EBRD documents. Hughes preserved its credibility with the Bank. OTE maintained its relationship with the EBRD on other projects for which it had received EBRD loans. UW continued operations and constantly begged for new investment.

Hughes maintained their credibility despite the following facts: the payment of the $5.0 M was financially ruinous to the business; their announcement several weeks later that they were out of the fixed wireless business was damaging; the complete failure of the MSU's destroyed the business plan; the certification of the system for only half of its capacity was ruinous. And where was our largest Shareholder, OTE?

With the EBRD out of the picture, we were left with no leverage. Hatwave owned only 48% of Ukrainian Wave. Since OTE owned the majority of the shares of Hatwave, no solution was possible without the participation of OTE. OTE did not like the original deal, but they would not tell us how they wanted it restructured. OTE would not sell its shares or withdraw from the business. After the default, and for the next five years, we reasoned and pleaded for a solution, but OTE did nothing. Doing nothing meant the company continued to operate, but there was no way to develop the company beyond what it could fund internally within the limitations of the equipment available on the used market.

The two men at the EBRD who had vision and a clear understanding of the purpose of the EBRD and the value of the power of the EBRD, Dellarosia, the Chairman, and Mark Tomlinson had left the Bank. Carl Gage, who had worked so closely with us during the process of EBRD funding was gone. Mark Tomlinson had also left the Bank. The bank was now safely in the hands of the bankers, who had no interest in the "Development and Reconstruction" mission of the Bank. It became merely another European bank. If the leadership that understood the mission of economic development in Ukraine had remained, the outcome for Ukrainian Wave probably would have been different.

People do make a difference.

My Last Story About Kirpa

In early 2000, during my last visit to Ukraine, Kathy and I met with Kirpa and his wife in his office. Our Managing Director O. Bilichenko, translated. I brought from my home a handsome large enameled railroad lantern, one of my favorite antiques. Kirpa gave us a large, beautifully carved wood platter, which I am now looking at in my living room. We said some personal and meaningful words about each other.

A few years later, Kirpa became the Minister of all transportation in Ukraine. In 2005, Kuchma could not run for a third term as President. Putin actively supported his chosen successor, Yanukovich from Eastern Ukraine. Yuvchenko had been the head of the National Bank of Ukraine, was familiar with foreign investments in Ukraine, and was pro-West. After the first election, a dispute developed as to the fairness of the first election. A mass strike developed and Kiev was closed down. Thus was born the "Orange Revolution" which supported Yuvchenko. We watched and read everything we could. One evening on the news, there was a picture of government officials. There, for a second, I clearly saw Kirpa. Because of the "Orange Revolution", it was agreed there would be a second election. After the first election, we were told that Kirpa was beaten up because he had not suppressed enough votes of the democrats in Western Ukraine.

After the second election, the pro-western Yuvshenko was elected President of the Country. Kuchma's chosen successor had lost. Several days after the election, Kuchma's Minister of Transportation, Kirpa, was shot. Of course the official word was suicide. I am still shocked and saddened. The method I was familiar with was car accident.

The Orange Revolution took on personal significance.

2005 The End of Hope
O. Bilichenko grew the company to 13,000 customers which generated income. But she became increasingly concerned as she reported at the end of 2004, 2005, and

2006, that they were reaching the limits of both the technology and the market. The engineers had begun to develop alternative plans, and came up with a plan for what they called Triple Play which they were able to test in the market. The company was begging for new investors.

In March 2005, O. Bilichenko and her chief engineer visited us in Cleveland. In 2005, O. Bilichenko had begun to develop a plan to purchase the Hughes equity and debt, in the hopes that she could buy out the western investors, Hughes and OTE, and obtain new investors. We could never find any investors who were willing to invest in the company as long as Hughes and OTE maintained their ownership interest.

In June 2005, Bill had surgery to remove a tumor, and in July he seemed cured. But in August, he began to have a temperature, which was symptomatic of the spread of the cancer. On October 5, 2005, we all received the most devastating blow of all. After several years of illness, Bill died. Kathy and I had said our farewell several days before. It was the saddest day of my life. Whether you met Bill and Linda in high school, college, work, or the communities where you lived, or were relatives, it did not matter. You were friends for life. No one could ever come close to their commitment to their friends—they traveled together, played together, ate together. There was always something going on. They were truly unique and inspiring. With him those last weeks were his sons, Bill Jr., Steve, and David, their wives and grandchildren, his friends and relatives, and of course Linda, and her sister. This is why, unfortunately, I am writing this without Bill's help.

Except for our last visit, Bill and I never talked about our friendship. We lived it. As business partners, we trusted and depended on each other. Together we were an extraordinary team.

In L'viv, the entire company attended a special candlelight Mass in Bill's honor.

O. Bilichenko sent the family the following message:

> "When you meet a great man, you know it from the first moment—you can see the wisdom of the world in his eyes, you can feel his strength, confidence and generosity. And that is the way we shall always remember Bill. He never ordered or lectured—he would give an advice, direction. He was the best boss - and mentor one could ever dream to have in one's life—I was really lucky to have had Bill as a mentor. He would show an example of hard work—when he was in L'viv, he would be the first to come to the office in the morning, and the last one to leave it in the evening. Who of the employees would dare to work less when they saw the Chairman work so hard? Everything he did had to be done perfectly—and he would re-write the business and financial plan over and over until it was ready to show to the investors. Negotiating with Ukrainian government officials was always a challenge—but Bill could carry the negotiations with such a grace, that even the nastiest bureaucrat eventually would do what Bill wanted—what was needed for the company. And it is only thanks to his guidance that over 10,000 families and small businesses in L'viv, Ukraine, and 120 villages around L'viv has modern telephone services. And we promise we shall do everything to implement his dream—to bring modern telephone services not only to L'viv, but to whole of the Western Ukraine.

But we also will remember Bill who enjoyed life—we had so much fun on the Halloween party he arranged for the employees!

He was also a wonderful, loving husband and father, and grandfather.

He was a great Man.

The candle that we lit today in the church of Saint George for Bill, will always burn in our hearts. Bill will always be in our hearts."

2007 The Takeover

O. Bilichenko, after a year's effort, finally completed the deal with Hughes for their equity and debt in Germantown Maryland at the end of May 2007. This gave her minority ownership in Hatwave, and control of the outstanding Hughes debt on the books of UW. But it did not give her control of Hatwave.

After completing the purchase from Hughes, O. Bilichenko and Kathy met in Washington D.C. in May 2007 for a brief talk about the next steps, now that she completed the deal with Hughes. O. Bilichenko wanted Communications Technologies to join her in asking OTE to sell their equity and debt, and she wanted to buy Communications Technologies' shares. Her idea was that Management would buy out the Western Investors and find new investors who would grow the company with new technology.

Then there was complete silence. In the fall of 2007, OTE contacted us, as they also had not heard from O. Bilichenko concerning a meeting she had scheduled with them. We had not heard from her since the meeting in DC. O. Bilichenko wrote finally that she was forced to sell everything and leave the business, because of her divorce from her husband, the second from the same man. The new owner would contact us. No news at all from anyone, and we began our own inquiries. The new owner claimed to own the entire company, and had installed his own Managing Director. Everything was done in secret. We received no information. OTE launched an investigation, through their Ukrainian lawyers.

2008 The Scheme

In May 2008, I received a copy of a document in Ukrainian from the Ukrainian lawyer hired by OTE. It purports to be a deed of transfer whereby Hatwave gives away all its shares to some unnamed party. I was expecting such a document. What I was not expecting was that this document was signed by Myron Stoll! There are at least five problems with this document, starting with the fact that I did not sign it. It is not my signature. There is no notary to my signature as required by Ukrainian law. It is only in Ukrainian without an English translation as required by all agreements. I have never seen the original and I do not have any translation, so I do not know what it says. Third, we do not even know who it is that now claims to own the company! Fourth, Communications Technologies is a minority owner of Hatwave, and I had no power, authority, or right to sell, much less give away, the shares of Hatwave. Fifth, and finally, who would ever believe I would give away the company after all these years. The question is now in the Ukrainian Court, and the question is whether the Ukrainian Court will protect a western investor, like OTE and CT, or allow a new Ukrainian owner to take over the company. Also at stake is the ownership of the Ukrainian partners in the Joint Venture, Ukrtelecom and the State Property Fund.

At stake are 13,000 customers, an operating company, licenses, and phone numbers and leased and owned property, a company worth $2.5 M according to the Ukrainian press. And at stake are the capable employees who have worked to develop this company.

This is one mountain I will not be able to climb. We no longer have the financial ability to do anything to protect our interests, nor the time, money, and effort necessary to protect our interests

It is as though the curtain came down in the middle of the play, stopping it, mid-sentence, mid-plot, amid hopes and dreams. The years of struggle to start the business suddenly became the play, not the prelude. But perhaps there was a play within the play. The Chinese have a proverb: Life is a journey, not a destination.

Our company, our employees, and our management had grown the company to 13,000 customers. There are and where people in L'viv, Kiev, London, and the US who believed in the vision, mission, strategy, and principles of the two guys from Cleveland and who believed the actual creation of the company was magical.

Ukrainain Wave and Hughes Employees

All the UW employees were exceptional. They were all dedicated to the business, worked hard, and loyal. All deserve mention. I will mention a few. Oles Kryschiv helped us from the beginning. Walter knew Oles from the Boy Scouts. He was an entrepreneur, and wanted to be part of the business. Oles always wore blue jeans, a bulge in back where he kept his gun, and an open sport shirt. Once the investors started coming to L'viv, I knew this was not going to work. I asked Walter to talk with him and see if he could look more like a businessman. Several days later we had a meeting which Oles attended. When he walked in I could not believe my eyes. Oles was neatly shaven, white shirt and tie, and wearing a blue suit. And he was carrying a black leather brief case just like us. He just looked at me and smiled. I never again saw him out of uniform. His company ended up owning shares in UW. Oles did a lot of good work for us.

Olya Onyshko helped us with translation. She was the daughter of a lawyer who helped us in the beginning. During 1992, when we went to L'viv for Ohio Bell, we always stayed in her Mother's home. She cooked for us and took care of us like a mother. One day I was sick as a dog, and could not get out of bed. She brought me a bag filled with stuff that smelled awful, and said put this to your nose and mouth all day! By the next day I was cured. Several years later, Olya spent several months with us in Cleveland.

Several years later she worked for an American company in Kiev. In 2007, Olya recognized Kathy in the ladies room at a museum in Washington DC. Kathy was astonished. Olya and her mother rushed out to the car to see me. Olya is living in Washington DC with her husband, a banker, and their two bright children. She was studying film making, and planning a documentary on Ukraine.

Our engineers were bright, young college graduates. Jerry Merkelo made one exception to this rule, our Director of Security. On the job training would not work for this position.

Jerry hired Dachko who was a former member of the KGB! He knew the system, and he knew the people. Most importantly, he was honest, great integrity, a good and respected man. He was tall, strong, and rarely smiled. He was a quiet, serious man. Sadly he passed away in the fall of 1999. The following was written at the time in his honor by Kathy:

Ode To Dachko

Dachko was a very "specific" person, as they say in Ukraine. He did things his way, and those around him were forced to respect him. He had a very intimidating presence.

As Chief of Security, he was in charge of everyone and everything. He hired the drivers and he selected the cars. He hired the security staff. He interviewed all job candidates and researched them thoroughly. He did not hesitate to add his own opinions and they were not without merit. He kept every legal document in a safe, and you had to have his permission even to peruse a specific document—but search for a missing document was almost unheard of. He trusted no one. He was also enormously loyal and committed.

When the equipment arrived, without papers, he somehow managed to get a faxed copy that met the government requirements. He loaded the equipment on a truck, without benefit of sides or ropes, and proceeded at 5 miles per hour through the city. When they came into an intersection and nearly collided with a trolley, and stopped all traffic, he simply got out and yelled a few choice words and everyone returned to their cars to wait. He hired a group of weight lifters to deliver the equipment, and it took 8 men to carry the heaviest of the equipment up seven flights of stairs and across the pipes to the installation location. The weight lifters strained to make this happen. Lesser men than

Dachko could never have carried this off. When he became ill, he still came to work, looking worse each day, but he was absolutely committed to the success of this venture."

Our role with the employees was to set an example, mentor and support them. They all wanted to grow and learn the business. They saw the value of team work, and solved problems by working together. In many respects, the employees gained the most from this venture, because many of them stayed with the company for many years and during that time they earned a good salary, and day to day working environment allowed them considerable participation in the operation of the company. What they never understood was the structure of the investment and the importance of generating some return to the investors.

Our engineers were particularly good. The chief engineer was Ihor Ronchevitz, and three of the other engineers whom we came to know well were Slavko, Roman, and Max. Hughes maintained that these were the best group they had ever trained. They participated in the designing the central office with the switch, the cell sites, and the network. They increased the capacity of the system first using second hand equipment from other Hughes sites, and later using fiber optics, another switch, and other software that served 13,000 customers in L'viv. With new investment, they would have expanded the system from telephony to "Triple play", using new technology. Ultimately, the engineers of UW made the company as successful as it was by their ingenuity and their clear understanding of the system gained over the years of operation.

Ihor I knew for several years, and he was a competent chief engineer, but he was a tough man, not a people person.

Slavko was tall, handsome, smart, and fluent in English, a people person. He was active in the Boy Scouts. We took many walks together. He visited our apartment many time along with other employees. Roman was shy, but an excellent engineer. We lost him when his wife won the lottery for a green card, and he is now working in California. Max was the person who developed the plans for expansion. He was smart, creative, and committed to the business. Once, with Olha Bilichenko, he visited us in Cleveland.

In 1999, we hired Ivan Griga as Director of Marketing. Ivan was an excellent Marketing Director. He was young, tall, handsome, a real people person with lots of energy. He put a good marketing/sales group together. Ivan was fun to be with, and we spent many hours talking with him. He left within a year when his wife wanted to attend college in Kiev.

There were two Hughes employees who played a key role in the planning and installation of the system, Jud Kenney and Wheeler Ramey. Both had worked on Hughes projects in other countries. They were knowledgeable, hard working, and truly good guys. We spent a lot of time with both of them.

Walking Back The Cat

I begin this summary by listing the things I believe we did right, and then listing the things that damaged the business.

Things that helped start the business.

1. Walter Bazarko was the reason why we went to Ukraine, and his friends were our early "boots on the ground" that established our relationship with

L'viv and Ukraine that was necessary to start the business.

2. Peat Marwick, a major accounting firm, gave their early support which was indispensable in enabling us to find and gain the financial support of the European Bank for Reconstruction and Development.

3. The international law firm of Jones Day took us on as a client in Cleveland, Ohio. We were then reassigned to Karl Herold in charge of their office in Frankfurt. Karl was indispensable in creating all the legal documents, both the financing and. corporate documents. Without his help, an international telephone company could not have been created.

4. It was important to have the support from the European Bank for Reconstruction and Development (EBRD). The EBRD provided all the debt capital required to start the business, and to grow the business. Obviously this is difficult for a start-up company to obtain. But there was a second powerful benefit from the involvement of the EBRD: The EBRD gave our investors and the national and local governments in Ukraine confidence in our business. Also never to be forgotten, the involvement of the EBRD gave protection to the company and to us personally.

5. The early support of Peat Marwick, Jones Day, and the EBRD gave credibility to our efforts to start our company.

6. We were fortunate with the support we received from the local, regional, and national governments in Ukraine. They helped us to overcome obstacles in Ukraine and obtain the necessary licenses and local support necessary to start the business.

7. We also needed the support of the United States government. The help we received from Ambassador Miller and the US Commercial Attaché Andy Bihun were not only extraordinary, but critical. Our company was one of ten companies on the Kuchma Gore list of US companies operating in Ukraine to be monitored by both governments. This support was indispensable.

8. We received support of the governmental and business community in L'viv. Our boots on the ground were all from L'viv, and they worked tirelessly and successfully in our behalf.

9. The support from Jerry Merkelo, the first Managing Director was critical. Jerry was born in Ukraine, spoke five languages, and retired from Hewlett Packard doing international business. He helped us from the beginning, and hired the best and the brightest employees. We also received significant support from out other Managing Director, V. Kalynyuk.

10. Thankful for all the employees. The Managing Directors, the engineers, the marketing and sales staff, the book-keepers and financial manager, the software and billing engineers, and the security staff were all honest, smart, hard-working and loyal to the company. Any weak link in these resources could have easily destroyed the company. It is extraordinary that there were no weak links.

11. The support from the Ukrainian Shareholders, the State Property Fund and Ukrtelecom, and the Railroad was vital. We obtained space from Ukrtelecom and the Railroad, and they were helpful in enabling us to meet changing legal requirements in Ukraine.

12. Kirpa, the head of the Railroad for Western Ukraine, was a powerful man who gave us powerful support. It was my plan, never realized, to use his railroad right of way to expand the company to Western Ukraine. The railroad, its employees, and its customers could have become good customers of our phone company. The railroad was a strong, strategic partner. Unfortunately, and to my constant amazement, not one of our investors had any interest in the Railroad, or Kirpa.

13. We were thankful for the support of our shareholders, OTE and Hughes who brought expertise in operating and manufacturing a telephone system.

Finally, and perhaps most importantly, support we received around the issue of bribery and corruption.

Since we started with the local government officials in L'viv, and the local people who helped and worked on our behalf, the word quickly got out that we would not bribe anyone. As related in the story, there was only one direct attempt, which failed.

Of course this is not an issue that is openly and honestly discussed. It should be.

There is corruption. But name me a country where that can't be said. More corrosive is the statement that one cannot do business in Ukraine (or any other country) without paying a bribe is not true, and often the excuse and justification for doing so is "they made me do it." Great. There is no basis for this holier than you attitude. It is my opinion that some companies are willing to do this because it is quicker than obtaining what is needed the old fashioned way, hard work. For some, this is an easy and sometimes cheap way to obtaining the objective. To be fair, paying certain "fees" sometimes requires honest judgment. Some payments can be legitimate because there is no money or structure to perform the services legitimately required. I do not believe, in appropriate circumstances, that is bribery. This is further complicated by the fact that most government officials, even at high level, were paid less than $60 per month. So you know the job is grossly underpaid for the power and responsibility they have, and thus the temptation.

The government officials I dealt with in Kiev and L'viv were competent and straight forward in dealing with me. No one ever suggested a bribe. It is true that serious and devastating obstacles were placed in our way. But I could see it in their eyes, and in their dealings with us, they wanted to see an honest business succeed. They really wanted us to succeed. One of the saddest parts of this story is did I let them down?

What damaged the company that led to its ultimate failure as an investment?

1. The time spent on raising the capital. In January of 1995, we filed our application for financing with the EBRD and concept clearance was given in June 1995. All the financing documents, however, were not completed and the first equity and first tranche of the EBRD loan were not paid until December of 1998, almost four years later. It was not until December 1999 that the equipment was shipped and installed, so that we could open the business to customers. We lost the advantage of being first in the market in L'viv and Western Ukraine, to two

national mobile telephone providers. Granted this was a start-up company, in a complicated, competitive business, but I honestly do not think it needed to take four years. At the EBRD, we started on the fast track and ended on the slow track. Underneath this over-arching delay, were delays caused by OTE's refusal to sign the Loan Guarantee and delays caused when Ukraine first voided all the licenses and then stopped our project again in 1998.

2. The critical damage was done when in early 2000, the HNS system and the E-TDMA technology was certified as capable of serving only 7,500 customers, not the 15,000 on which our contract and the loan and the investment were all based. This substantially increased the initial capital required. The lack of MSUs to serve large clusters of customers in apartments and lack of portability also limited our ability to economically serve the market. It is important to put this in perspective. In the 1990's, the TDMA technology was deployed in many countries. There was a lively debate as to whether CDMA technology would be commercially viable. During this time when the Hughes E-TDMA technology was selected, doubling the capacity of a TDMA system, it had been installed in several countries and Hughes won an award in 1996 for the technology. We were not aware of any experts at the time who said or predicted the system would not double the capacity of TDMA technology.

3. Finally, the Master Supply contract with Hughes had required that they be paid the initial payment after the equipment had been installed and accepted by Ukrainian Wave. Certification by Ukraine government was the responsibility of Ukrainian

Wave. Hughes insisted on this condition, but it turned out to be a big mistake. The remaining Hughes debt was second only to the EBRD.

4. This led in 2000 to the perfect storm: created by regime changes. At the EBRD, there had been a total change in leadership and there was no interest in saving the company or development in Ukraine. In January 2000, Hughes announced it was out of the wireless telephone business. By that time, it was no longer owned by either Hughes Electronics or General Motors—it had been sold to Direct TV. At OTE International, there had been a complete change in the leadership, and strategy as they sold their interests in other national telephone companies in Armenia and Romania. With the EBRD out and Hughes out, OTE's new management saw no value in continued investment in a small private company, like Ukrainian Wave.

Since we had never been paid under our Management Contract, and had lost so much of our personal investment, we could not afford to go back to Ukraine, so we no longer had our own boots on the ground. We did continue, however, to keep in constant contact with O. Bilichenko. We did try to find new investors, but none were willing to take on the challenges of old investors and new technology. Management kept up the operations, but ultimately it had to find an alternative for the company to remain competitive in what was a rapidly changing market.

What we learned from starting our business in L'viv, Western Ukraine.

We encountered too many potential investors who saw doing business in Ukraine as a risk for investment, as opposed to an opportunity—a sizable market—to provide services to a whole region of Ukraine, once established in the major city in the area. This would give us access to a potential customer pool of 10 Million people. Ukraine had not established itself as part of the global economy, although it was eager to participate, but we encountered few outside Ukraine who saw this as an opportunity.

There are now large and small businesses participating in the global economy and that makes more opportunities for American business to grow within the global economy, not just in the US. We saw an example in Cleveland. A lamp shade company hired a recent college graduate, a Chinese woman. She urged her company to do business in China, and told us after several years, 40% of the company's business was in China.

Since our business was located in Ukraine, we had to learn new things, meet new and different people, and understand their history, their culture, and their legal systems. The Ukrainians saw and understood we were committed to them. We stayed in a hotel where Ukrainians stayed. We had an apartment in their community. We went with them to see and understand their history, their culture, their art and music, and their personal events like weddings and birthday celebrations and religious services. From these efforts, both sides grew trust and respect. The significant result was a shared commitment to positively meld our way of doing business with their way of doing business. In the end, it was no different to do business in the complex and diverse society of L'viv than it is learning how to do business in the complex society of Chicago, New York, or Cleveland. A business man in any community faces many of the same problems and the same need to find relevant and creative solutions.

This was our business. We were in it together. This bottom up approach made the effort interesting and certainly exciting. To make this effort was not a daunting challenge or a pain to be endured, but a privilege for both sides to learn new things, meet new people, to form new friendships. Both sides aspired to build a new, successful, and profitable business. I would argue the bottom up approach is the key to success in any community. This I hope is the enduring lesson of this story.

What must be constantly applied is the American work ethic, the American creativity, American honesty, and the American way of finding collaborative solutions. What does need to change is that Americans not just go to cities in the world as tourists, but as people going to do business. We cannot rely on our economic or military power. We have to rely on American values. These values make the difference. Bribing people in the long run is not a short cut for hard work.

Finally, what is of critical importance to the international entrepreneur is the help, support, and assistance of the International Development Banks, like the EBRD in London and the Western NIS (Newly Independent States) located in Washington, and the help and support of the US Department of Commerce and the US Ambassadors. Money cannot buy what they can deliver. Both the Banks and the government must strengthen their focus on development and the support of the entrepreneur. As the reader already knows, we would not have been funded without the commitment and support from the US Director on the EBRD Board (we met with him several times), the bankers, Mark Tomlinson and Carl Gage, the chief of staff, Ron Freeman, and personal commitment of Dellarosia, the Chairman of the Board. That is a lot of support!

But in 2000, when the situation turned difficult, there was a complete change. The US Director to the EBRD sadly had died in a plane crash, and the key bankers all left. We not only lost their support, but we lost the commitment and courage to help us find a solution. We urgently sought that support, but none was given. If the Vision is gone, and the strong sense of Mission is not there, then there will never be any meaningful and substantial support. The Bank was now firmly in the hands of the bankers, and bureaucrats, who did not focus on development. With the loss of the Bank's support, we lost the investors' support. To support our position that the company could and should have been saved by refinancing and a new business plan, even with the loss of the support from the Bank and our two major investors, over the next six years the company grew to serve 13,000 customers, using internally generated funds and their technical expertise to expand the system.

If this business we started was not worth saving, let me conclude with the following story. During 1999, Bill and I had several meetings. At one of the meetings, I arrived early. A person already there began telling me that Bill and I did not deserve to run the company. He continued by saying, "All you know is L'viv. You do not know Ukraine. In fact you do not even know where Uzhorod and Muchacovo are." I was stunned. There were many options to answer these statements. In anger, I simply said do you want me to pick up the phone (not in quite such polite words) to inform you how many times I had been in both cities. Forget what Bill and I had accomplished in Kiev.

Three different people, knowledgeable about our business tried to put a group together to take over the company, without our knowledge or participation. Of course, the final successful effort was set up by our own Managing Director. There seems to be no bounds on arrogance or jealousy— take your pick. But they would have been happy to take over the results of our work.

And ultimately, someone did.

We hired O. Bilichenko as the Managing Director in the last quarter of 1999 when the installation of the equipment was nearing completion. This was the time of maximum hope and opportunity. Within eight months, everything changed. The limitations of the system as determined by the certification, the lack of additional capital, default, and a dysfunctional shareholders group were devastating blows. She did not jump ship. She did not let the company go out of business. She kept the employees who could help the company survive and succeed. By sheer hard work, intelligence, and skill, she and the employees grew the company to 13,000 customers. She managed to buy additional equipment and additional single subscriber units. Bill continued to serve as Chairman of the Board, even though there were no Board meetings. Bill, Kathy, or I were in weekly contact with her for the next seven years. We had a close business and personal relationship with her. Without her leadership, there would be no company to take over. And thanks to her hard work, leadership, and commitment, the employees of the company have had ten years of employment in which they have learned much. They will be an important asset to any company, including the new owner.

All we know now from her about what has happened since

May of 2007 is that she immediately sold her shares which she had just acquired from Hughes. We believe that she had had to borrow a substantial amount of money to buy the Hughes shares and debt, because it had to be paid in cash. She told us she would take a mortgage on her house, but we do not know what that means. Was it a bank loan or a personal loan? We do not know whether she was forced into this or why she cannot or did not tell us more. In 1999, Ukrainian Wave was required by the government to install a monitoring system, so the government could monitor all communications. She changed her email address, but that too can be monitored, as well we know. Perhaps she was afraid to communicate with us from Ukraine, but then what did she know when she met Kathy in 2007 in Washington D.C., just weeks before she sold the shares and debt she had acquired from Hughes. It is a mystery and a significant loss.

The factors which justified our decision to start our business in L'viv included the following:

1. Ukraine is a large country with nearly 50 million people. It took the people over 1,000 years to become a free, independent, and democratic country, which they finally became in 1991. Ukrainians are justly proud of this historic struggle.
2. Since Western Ukraine had been dominated by Western Powers, their architecture is Western, and the region is the most fiercely independent and democratic. Eastern Ukraine for much of its history was dominated by Russia. This historical split has influenced the split which still exists today between the Western and Eastern parts of Ukraine.
3. Since Western Ukraine is the cross roads between

North, South, East, and West, they have a long tradition as traders and businessmen. They are also well educated and smart. L'viv University is the oldest University in Ukraine. We saw a strong desire to meet and do business with foreign businessmen. What impressed me the most was the respect for Americans!

4. With an improved infrastructure, their rich black soil could support a strong agriculture.

5. Ukraine has made a determined effort to restore their historic buildings and churches, and to preserve their cultural heritage.

Finally, but not least, the young people are smart, well educated, and creative.

Final Thoughts

In the end we lost our money and our investors lost their money. The sole purpose of this effort was to make this a financial success, not a meaningful journey. Every effort from beginning to end was dedicated to the single purpose of financial success. However, we also knew that life is a risk, and entrepreneurship is by definition a risk.

Especially to Bill's family, and to all of us, the real loss was losing Bill. He is irreplaceable.

We did learn that people do make a difference, and in this journey people did make a difference. A lot of people trusted each other. We understood our limitations. We did not try to impose our limitations on others, but to open them up to new possibilities. We touched and influenced each others lives. Most of these relationships were intense, sustained over many years, and were personal, challenging, and exciting.

Ukraine is a beautiful country. The people are extraordinary. All people are examples of the true meaning of the assent of man. "The Ascent of Man" was written by J. Bronowski. His book affirms the everlasting value and meaning of the growth of the human spirit. Human values, beliefs, and intelligent striving mean everything.

I am sad that the project we started, the hopes and aspirations we ignited were not directly fulfilled. Dreams, hopes, and ideas do live on.

My hope and confidence is that all of us will believe there is no disgrace in trying. Who will ever know if we made a difference. But without trying, there will never be a difference.

Of course the purpose of this journey was to start a business. Our intent and belief was we would make this business a financial success. Often in one's life, defeat is just as fleeting as success. Often the journey is life's greatest gift. If I had the choice, yes I would do it again.

Our world is overwhelmed with opportunities. Believe in your own future and believe in your passion to take advantage of those world-wide opportunities.

Now I circle back to the beginning, the Preface. The reader can now understand why the families were so moved by what Karl and Lute wrote.

I end to go further than I began. My vision for Americans today is to go everywhere in the world as entrepreneurs. More people today may be ready to join the Dreamers and ask why not?

L'viv Opera House. On Left: Myron and Bill. On right: S. Dembicky, Wheeler Ramey, Hughes, second from right.

M. Stoll, Downtown L'viv.

K. Stoll and J. Merkelo, Managing Director, Ukrainian Cultural Garden, Cleveland.

Part VII: Ukraine, L'Viv, the Times, and the People

Ukraine

Ukraine is part of the largest flat land mass in the world, extending from the Baltic Sea through Eastern Europe to the Black sea and across Russia to the Ural Mountains. Geography is destiny. Armies from empires and nation states easily marched through and controlled much of this land mass. Like the tides of an ocean, for over 3000 years, armies conquered, lost, conquered again, and lost again. At one time, in the area of Ukraine, there were eleven different armies marching in and through the territory that would become Ukraine.

In the area near Kiev, people had settled in the Neolithic period, before 4500 B.C. About 700 B.C. the Scythians and Sarmatians came to the area. Then ancient Greece and Rome each had their time of influence, as I saw in the cultural evidence in a Museum in Kiev. From the 13th century to the 20th century, the territory was occupied by the empires of Lithuania, Poland, Turks, the Ottoman Empire, the Austro-Hungarian Empire, Russia, and of course, Germany.

The golden age of Kiev and the real beginnings of Ukraine spanned the 9th-11th centuries, when the people known as the Kievan Rus were in power. By the 11th century, they were the most powerful and significant group in Europe. These people included Varangians from Scandinavia, who had become assimilated with the local Slavic population, and were known as the Kievian Rus. After the 11th century, the area was dominated by the foreign empires as cited above. In the 17th Century, the famous Cossack military, striving for independence and freedom, fought the Turks, the Tartars, then Poland and were finally betrayed by Russia.

Christianity played an essential part in the development of Ukrainian culture, in the form of the Ukrainian Greek Catholic Church, the Orthodox Church, and the Roman Catholic Church.

The tradition of the Easter Egg goes back to the Slavic people, thousands of years ago, with its beautiful patterns and bright colors. The tradition of the eggs is an important part of Ukrainian culture, with each region having its own tradition.

With its rich black soil, agriculture has always played an essential role in Ukraine's economy. When Ukraine became part of the Soviet Union, minerals, mining, and manufacturing also played an essential role in the Soviet Union's economy.

On December 1, 1991, a referendum was held in Ukraine and over 90% of the people voted for its independence. Leonid Kravchuk was elected its first president. It became one of the Newly Independent States (NIS). Ukrainian became the official language for the first time. In terms of size, Ukraine is the second largest country in Europe, second only to Russia, with a population of fifty million. Ukraine is over 600,000 square miles, slightly smaller than

Texas. It is bounded by Belarus, Hungary, Moldova, Poland, Romania, Russia and Slovakia, and the Black Sea and Sea of Azov in the South. The human capital of this country—the people—were smart, well-educated, and understood human nature.

Two catastrophes have occurred in Ukraine. In the early 30's during Stalin's control of Ukraine, as result of the process of collectivization of farms, a terrible famine took the lives of military and peasant farmers, in which 8.0M people died—more than one tenth of the population. Other policies imposed resulted in the burning of books and the murder of educated people who were thought to be subversive or worse. Virtually no one was untouched by the oppression of Soviet rule. The second catastrophe was the devastating Chernobyl nuclear plant explosion, which caused birth defects and compromised the health of its young people.

This chaotic history molded the language and the culture. Ukraine is a very specific country. That is, things must be done in a very specific way. The Ukrainian language has often been repressed by those who occupied the country and the Ukrainian language had not evolved as the language of law or governance, even in Ukraine. If anything, it was the language of longing for freedom, of protecting family safety. During Soviet occupation, Russian was used as the language of law and governance. Ukrainian bears vestiges of Lithuanian occupation, for example the use of the term Hetman, literally the head man in a given area. But more important, the vocabulary in Ukrainian was not fully developed for legal or business purposes. For example the same word in Ukrainian can mean the bill at a restaurant, the profit of a business, or the winnings from a lottery. So translating an English legal document that used such words as deliver, implement, carry out, and execute was redundant in Ukrainian vocabulary. The same Ukrainian word would be used for all these purposes. There was no word for the concept of "gender."

Another specific thing about doing business in Ukraine is that its historical legal system is based on Roman law, not English Common Law. Thus the governing authority is vested in a person or position, and any binding document must be signed in person, properly signed, and stamped, with adequate evidence of what is to be provided to the notary. Only a stamped and signed and notarized document has legal standing, and in many cases a multiple page document must be sewn together, to prevent any modification of the original as it is transported from one person to another. And further, the original documents must be secured at all times. If there is a dispute over a contract, the notary is the only source of evidence acceptable to the court.

For all these historical reasons, the people are very protective of their person and their family. One never saw family pictures or personal items of any kind in any office, although it might be furnished well, with oriental rugs on the floor. It impressed us as very formal hierarchical social system based on position and authority. For example, at all of our meetings, the Ukrainians sat on one side of the table and the "westerners" on the other. When we tried to mix it up, by changing sides, or sitting on both sides, it somehow ended up in this formal arrangement. The people are well-educated (99% literacy rate) and frugal. Although there are buildings that speak of historical wealth, people did not show their wealth, even if they had it.

L'Viv

L'viv is a city of 800,000 people, the largest city in Western Ukraine, and regarded as the second Capital of Ukraine, according to some of our Ukrainian friends. A Prince who named it after his son Lev, which in Ukrainian means Lion, founded the city in 1265. The City celebrated its 750th anniversary in 2006.

In the 14th Century, it was the important trade center for the Black Sea and the Baltic trade route. From the 14th Century to the mid-18th Century, Western Ukraine was under the control of Lithuania and then Poland. From 1772, until the end of World War I, the city was part of the Austro-Hungarian Empire. Ukraine was finally incorporated as a nation and became part of Soviet Russia. There was no reason to bomb L'viv; it was not a center of industry or mining. But the Germans moved trainloads of food and black soil and people to work in their factories in Germany.

Even though L'viv for almost a millennium had been under the power of foreign states, the City kept a unique spirit of open-heartedness, diversity, democratic traditions and always a desire to gain freedom. The City has always maintained openness, a cosmopolitan and commercial identity. To the City came Hungarians, Cossacks, Russians, Turks, Poles, Armenians, Jews, Roman Catholics, and Greek and Russian Orthodox. This was reflected in the names for the City—Levensburg, Lemburg, Leopolis, L'vov, and L'viv. This was also reflected in its architecture: Gothic, Baroque, Renaissance, Roman, Rococo, and traditional Ukrainian. It is on the western edge of the country, 100 km from the Carpathian Mountains, and 70 km from Poland. It is 280 miles from Kiev. The City averages 296 m above sea level, with its highest hill 409 m. above sea level. The River Poltava goes right through the City. In the 20th Century, the river in the center of town was covered by the Mall. The rest of the City is built on hills. This is not insignificant when planning a radio-based telephone system.

During the Second World War, the Germans did in L'viv, what they had done elsewhere, virtually destroying the active Jewish community and most of its 45 synagogues and sending the Polish residents back to Poland. The Armenian Church is one of the few vestiges of that diverse community. L'viv has often been the center of intellectual activity and political activity, and it was here that the first efforts at building an independent Ukraine began. Through it all, the City developed and maintained its rich culture and spirit. The City was always in the center for the preservation of Ukrainian culture, architecture, and a democratic independent Ukraine.

Many of the existing buildings and churches were built in the 16th-19th Century. L'viv University, the first in Ukraine, was founded in 1784. In the center of the Rinok Square, a perfect European city square, is City Hall. In the center of the City is a four block open mall, where people gather to talk, play chess, and sing. At one end of the mall is the magnificent L'viv Opera House, which is reputed to be the second most beautiful Opera House in Europe. We saw wonderful performances of Rigoletto, Madame Butterfly, as well as Swan Lake, and several Ukrainian ballets—one a very old story of a Hetman and another contemporary ballet. The Opera House had a resident staff of musicians, dancers, and performers of 450, and almost daily performances.

It is clear that music is a central part of the culture of Ukraine. In addition to the L'viv Opera House, there is also the municipal concert hall. I heard many outstanding recitals and chamber concerts there. I heard children recitals. I still remember a concert by a Russian oboist. He was breathtaking, although I never heard him take a single breath. I heard people sing in the church choir, in the public parks in the evenings, and in bars and restaurants. I heard outstanding concerts in Kiev. Both music and the Catholic religion are strong parts of their culture.

The first train arrived at the newly built L'viv Railway station in 1861. In 1884, the city established a telephone communications system, and the first electric street cars began in 1894. Today the Railroad connects nine major routes, including connections to Slovakia, Budapest and Poland. The L'viv airport had no international flights when we began to travel to Ukraine. One had to fly from Frankfurt to Kiev and from Kiev either fly or take a train to L'viv. Today, there are direct flights to L'viv. In the late 1990's, Kiev built a modern airport. Good public transportation within the city and good connections to Western Europe continue to be major assets of this area. It continues its role as a market center for Western Ukraine and connections north and south, east and west.

The first two decades of Ukrainian Independence have yielded advances for the Ukrainian people, their rights, their language, their economy, and their legal system. But the political situation is still difficult. Several people have "committed suicide", one journalist has been killed, and the past President had been poisoned.

Industries were privatized in Ukraine, and many "oligarchs" have made some serious money. These "oligarchs" are viewed with great suspicion on the assumption that they acquired their money by unlawful activities and or political connections. In reviewing the public information available about five of them, it is apparent that they were positioned in industries that they knew and when the mining and manufacturing and energy industries were privatized, they already had some ownership and knew how to make the companies successful. They did not have cash. In fact, the early development involved bartering not only between themselves, but with some Russian companies. They did not borrow from international banks, and foreign investment was limited. Their economy was home-grown without some of the stringent "reforms" imposed on other countries like Poland and Russia or by international banks or foreign investors. They protected their assets, they protected the local labor market, and the economy has grown. Now some of the oligarchs, for all the questions about their unlawful or political activities or their extreme wealth, are beginning to invest in other industries as venture capitalists or in philanthropies of their personal interest, such as AIDS or government reform. Some are active in political parties, and some hold political office. They did not benefit so much from external investment by international banks or foreign investors. They made their local resources work.

In this respect, one could argue that they are not too different from some of our historic venture capitalists, like Mellon, Rockefeller or Carnegie, who restructured and consolidated their industries, moved on when their industry was restructured, and left a legacy of philanthropy which the USA still enjoys. Ukraine continues to have robust mining, manufacturing, and agricultural industries. Foreign investors are still at risk in Ukraine because the legal system does not protect them. It is an insider's game, but it is a Ukrainian game.

But the country is still losing population, and finding the human resources for some hard industries is difficult. The literacy rate is 99% so education is available and good. One of their thriving industries is software and information technologies.

The country has been divided on whether to enhance their relationship with Russia or the European Union. This is the fundamental issue that divides the population and the political parties. Eastern Ukrainians view Russian as their native language, and are more inclined to maintain their long relationship with Russia, and they control the major mining and manufacturing industries. Western Ukraine views Ukrainian as their native language, and wish to maintain their historical trading relationship with Europe. Crimea in the South on the Black Sea is an important area of the country, under Ukrainian control, but with certain independence and a nascent desire on the part of its Russian speaking population to again be part of Russia, as they were before 1954.

The most serious issue is Ukraine's membership in NATO, because of course that would damage the country's relationship with Russia. That poses many problems, since the country is dependent on Russia for trade and for natural gas. But membership in NATO would enhance the country's trading relationships with the European Union, a separate more independent path, on which the country seems to have some agreement. But the people are so divided in their political party allegiance it is hard to form a consensus. There are a large number of political parties. There are local elections to send local representatives to the

Rada or congress. Once V. Kalynyuk took me to a voting place to watch how people voted. There were lots of people in a crowded room. It appeared easy to vote. The national elections for President continue to be divisive. So in spite of the serious advances the country has made, the political situation must be described as unstable. Add to this story that Putin has actively supported the current President. The United States has imprisoned a former Prime Minister for money laundering, and refused a visa to another important political personality, and it is clear that Ukraine is still vulnerable to foreign interference in its affairs.

The Times

It is important to remember that we began this journey on year after Ukraine became an independent country. We had many fascinating experiences.

On my first flight from the Kiev airport, an incredible incident occurred. It was necessary to take all the passengers from the terminal to the plane by bus. All the passengers except me remained on the bus. An armed soldier escorted me to board the plane. Only after I was seated, were the passengers allowed on the plane. They did not know who I was. All they knew was that I held the telling dark blue American passport. This was to show their respect for Americans. While this custom quickly disappeared, what never disappeared was the respect Ukrainians held for America. That was why we felt we could do business there. In those early years, it was necessary to go from Kiev to L'viv by a 12-hour train ride. In Kiev, the luggage from the plane was taken by hay wagons to the terminal. Today, Borospol is a modern efficient airport.

On Bill's first visit to L'viv, we stayed at the Grand Hotel, directly facing the center of the mall. It was a bitter cold January night when we returned to the hotel about 11:00 P.M... Across the street in the center of the mall was a group of about 100 men and women singing Ukrainian songs. In the local park during the day, children came down the nearby snow covered hill on wooden sleighs, with high wooden runners. I never saw a parent walking with their children, regardless of the age, without holding hands. When we arrived back in Cleveland, we both bought and read Orest Subtelny's 600-page history of Ukraine. This is an outstanding book, thorough and well written.

Bill also met my first friend in Ukraine, Nicholas Sergienko and his wife Sveta. John Papay, an employee of Ohio Bell, was an active ham radio operator. The local club regularly talked with Nick in Kiev. They asked me to contact him, which I did. For two years, whenever we were in Kiev, we stayed with Nick and Sveta. Nick was a professor of Metallurgy at the University. Nick was a heavy set man, but in great shape. Countless times we walked all over Kiev, visiting the historic churches, museums, and historic buildings. His English was quite good. When he had questions, his favorite expression was "Mr. Myron, explain me please."

As one crosses the Dnipro River, entering historic Kiev, the first sight is a silver plated statue of the "Mother of all Mothers Victory," about 100 feet tall, holding the handle of a large sword. She is in a large park, a memorial to those who died in the war, and near the eternal flame. Next to the park are many of the historic churches.

One evening, we were walking near the restored golden gate, built in 1037 that was part of the wall that protected the old city. Nearby, a large building was on fire. Next to the building was a 1930's fire truck, complete with large white wall tires. Two firemen were on the ground trying to throw the hose to a fireman hanging out a second floor window. He could not reach the hose, which was stuck in a tree. It reminded me of a 1930's Charlie Chapman movie.

One afternoon, Nick treated me to a memorable event. A friend of Nick's was the Managing Director of the large municipal theater. About twenty people were there as they were filming the Red Army Chorus for French television. With the takes and retakes, the concert lasted several hours. It was as thrilling as it was spectacular. I videotaped about half the concert, until my battery died.

We had many good times with Nick, Sveta, and their son and daughter. On one occasion I presented Nick with an honorary plaque from the Cleveland Ham Operators Club. Sadly, as our project became more complex and time-consuming, I lost touch with the family. I never had the time. Several years later, I tried to re-establish contact, but was unsuccessful. Some thought that Nick had died. I deeply regretted that I had not maintained contact with him and his gracious family.

On one of my trips, I stayed in the Dniester Hotel in L'viv. On each floor, a woman sat behind a desk watching the floor. I gave her some white shirts to be ironed since the next day I would be taking the train to Kiev. I gave her a good tip to pass on to the cleaning woman who ironed my shirts. I always wore a pouch around my neck, tucked under my undershirt to carry my passport and money. It always amused people watching me as I partially undressed to get my money or my passport. On the train, I woke up in the middle of the night and did not feel the scratchy pouch around my neck. I panicked. In my haste to leave, I had left the pouch on my bed. When I returned to my hotel room after my trip, the pouch was not on the bed. I looked in the drawer, and there neatly rolled up was the pouch, all my money and my passport. In all my countless trips to Ukraine, no one ever attempted to steal anything and I was never threatened. While Bill and I met countless numbers of people, we strictly avoided any publicity, no radio, TV or newspaper publicity. All that would accomplish is to make us as a target. All publicity was confined to Ukrainian Wave and its employees.

Other than our friend from the U.S. who owned a business in L'viv, and the Hughes employees, and one consultant, we never met an American doing business in L'viv. This was a great disappointment. The Grand Hotel was originally purchased and renovated by an American, but she had left before we arrived.

In the early years, we would sometimes fly to Poland where a railroad driver would meet us and take us to Ukraine. At that time, there was only one border entry from the North. Obviously, the Soviets did not want to make this easy, and it was not. There was one long back country road connecting the two countries. The line of cars and trucks waiting to go through customs stretched for miles and miles. Generally this was not a problem for our driver. Our driver would make arrangements with someone from the Polish Railroad who would make arrangements with customs. Of course this was exciting. The long country road had one lane going South to Ukraine, and one lane going north to Poland. Don't ask me how he did it, but we drove most of the distance in the on-coming lane. Once, when communications became confused, we spent the entire night in the line. Today there is a new modern border crossing at a different location.

Living in Ukraine

During all the years we were in Ukraine, our only activity for recreation and exercise was walking. We walked through much of Kiev and most of L'viv. There were historic sites we could only see by car, but the majority of the time was spent walking. In the early years, our friends and security people were strongly opposed to our walking at night. But they quickly gave up. Our original office was about three blocks from our apartment. Our second office in the Railroad Building was close to the center of town, about fifteen blocks from the apartment. Unless it was too cold, we would walk home from dinner or the office to our apartment. We knew the city well. Except for the gypsy beggars, the people were neatly dressed and no one looked poor or unhealthy. The streets were always crowded with people. There were electric trolley cars throughout the City, but L'viv is a "walking city."

Early on, I was walking with one of our engineers. He said to me, "Everyone knows you're American." "How could that be," I said, "I am just wearing jeans and a shirt." "You're always smiling." he answered. It was true. When they are walking, their expression is serious and private. The Ukrainian people do have a good sense of humor, but not an Irish sense of humor.

The town was filled with many small shops, coffee shops, restaurants, and a few antique stores. At the south of downtown, there was a large open market. At the front end was a large flower shop. People bought a lot of flowers. Most of the market was open, but covered by a roof. There were rows of wooden tables for the sale of food. I saw lots of cabbage, beets, bread, and some onions, carrots, and potatoes. I saw chicken and a lot of pig parts, but very little beef. All the food including the meat was laid out on the wooden tables. There was no refrigeration or packaging. Kathy learned to wash food before she cooked it in vinegar and water.

By 1998, we had rented an apartment within walking distance of the office. Near our apartment was a deli. We bought most of our food and drinks form the deli. We bought rotisserie chicken some meat, cheese, bread, juice, beer, and vodka. The ladies in the shop knew us well, and always enjoyed our buying using sign language.

Our apartment included two large bedrooms, a large kitchen and dining room, and a small bathroom. The dining room with a large wooden table also served as our office. There was a beautiful painting of L'viv on the wall. There was a stone archway going from the kitchen to the front hall and bedrooms. One walked up one flight of stairs to enter the apartment, but the kitchen and dining room were almost ground level at the back of the building. There were bars over all the windows. As was true with most apartments, the halls were dimly lit and dirty. No one cleaned the hall, so Kathy bought a broom, and swept the front entrance and the hall. One could take a hot shower before 8:00 AM. After that the water was off or cold. The back door to the kitchen was nailed shut.

Often Bill or I would be there alone. I must admit there were times I was a little nervous. There was one occasion when the security of the apartment was put to the test. One evening Kathy and I took a cab home after dinner. As the cab left, we realized we both had left our "kluch" (the key) in the apartment. There was nothing I could do but go to the office to ask for help, leaving Kathy alone in the hall. I left in a panic. The security guard on duty was a young man I knew and liked. I explained that Kathy was at our apartment and we had no "kluch." The security guards were instructed never to leave the office unguarded. He grabbed me and we both ran back to the apartment. There was no way to get in the front door. It had a secure double lock. We walked to the back side of the building, and he looked up at the bars over the windows in our first floor apartment and smiled. He climbed up to the deck, grabbed two bars and bent them apart. In one minute he was in the apartment and opened the front door. While we were relieved to be in the apartment, it did not add much to our confidence in its security.

One night, Bill and I were drinking beer at the Grand Hotel. We heard a young man speaking English. Excitedly, I turned towards him and asked where did he live in America? He told me he was Australian, and owned the bottling company in L'viv for Coke.

One humorous event occurred when I was flying home for Christmas. After I presented my passport to the Custom official, I then went to check my bags. I was carrying two large paper bags filled with bottles of vodka. One of my suitcases was filled with bottles of vodka. Well, at $3 per bottle, and the holidays coming up, why not? The customs official looked at me, and then looked at the bags filled with vodka, then looked at me, and then in effect told me you can't do that. Now we all know each other since I was doing this almost every other month for years. I reached in and took a bottle of vodka, ran back through customs (not allowed) back to the lobby to find the Manager of the airport, (I knew him too), did so, and gave him the bottle. I went right past the custom official who checked the passports, back to the custom official checking my bags, with a big smile. She looked at me, tried to suppress a smile, shook her head and waived me on. I wanted to give her a bottle, but thought that might get us both in trouble.

In the Center of town, next to the mall leading to the opera house and across the street from the Grand Hotel was the Vienna Café. It served good food, and in the summer, there were many tables on the terrace outside the restaurant. It was a favorite resting place for foreigners. Over the years we got to know many people from other countries doing business in L'viv, but no Americans. Most summer evenings we sat outside drinking beer and eating dinner. The views of the center of town, the opera house, the tower of city hall, were so special and enjoyable. This ritual became our favorite pleasure.

One evening while eating I looked back towards the tower of city hall with a view of the street that was in front of Rinok Square. I could see with this view one building, and one half of the building on either side. All three were freshly painted. Now I was curious. Since this was only minutes away, I went to check it out. The center building to be seen from the view was completely painted. The half of the two buildings on either side that could also be seen through the view was also painted, but the rest of the two buildings were not painted. It turned out that many top officials from European countries were coming to L'viv for an important conference. You could tell what parts of the city they would see by where things were painted. For example, if you came to an intersection, if the black railings in front of the houses or buildings if you turned left were painted and the railings to the right were not, you know they would drive left.

Another reason we all liked the Vienna Café was because it had decent toilet facilities. Especially in the early years, this was a constant problem. One hotel we stayed in had a hole in the ground. There were not many public places like train stations, or private places like restaurants that had modern facilities, not holes in the floor.

About a block from the Vienna Café was an open flea market. I bought wool blankets, an exquisite flower painting for $40, and a handsome beautifully carved walking stick. The first or second year we were in L'viv, I bought at the Ukrainian Museum, which was in a magnificent baroque building, an antique coral necklace, with many long strands of red coral. I also bought antique linens and blouses, beautifully embroidered.

By the mid-nineties, none of these were available. In an antique shop, I bought a Rosenthal numbered plate. Back then, Ukrainian –Russian mink hats were also available. They were called Norkia Chapulkas'. The mink hats were handsome, warm, and the mink fur soft and fine. I had a sable mink hat made for Kathy. When I wore mine in the US, I could have sold 100.

Through a friend, I was able to buy a painting of the Black Sea that was extraordinary. It reminds me of the eternal ebb and flow of life. It was done by one of Ukraine's major artists, Koval, who specializes in seascapes. There are few artists that can paint better seascapes. A Russian painter and friend of Koval's painted the sky. It is a masterpiece. One cannot tell if the sun light coming through the dark clouds above the crashing waves is the end or the beginning of the storm. It now resides in my living room, near the Atlantic sea.

After Ukraine's Independence, the first available money was spent on restoring the historic buildings, especially churches. This involved restoring very beautiful ornate architectural details and the roofs with gold, a very expensive proposition. When we asked why they were spending so much money on churches instead of infrastructure like railroads or transportation, the answer was that first they had to restore their history. When you know the history of Ukraine, you can understand the importance of re-establishing their Ukrainian identity.

Kirpa arranged for me to visit the hospitals owned by the Railroad. It was a frightening experience. Sanitation and medical equipment were almost absent. This was because they did not have money to invest. This was sad, because the people were bright and knowledgeable. To put it in perspective, in 1996, when we were starting the company, the annual revenue of Hughes Network Systems was the same as the tax revenue for the entire country of Ukraine, with 50 Million people.

All the education, transportation, health and social services were provided by the government. Often these services were weak because of lack of money. From our experience, the strongest of these services was education. The Boy Scouts was the only voluntary association or non-profit organization we saw or heard of in Ukraine.

Over the years, I made countless trips to Kiev, the Capital, by train from L'viv, a distance of 280 miles. The trip took twelve hours, and I always had a two or four person sleeping car. I actually enjoyed the train ride. I enjoyed the constant clickety clack. Most of the time we traveled at night on the train, arriving in the early morning. Over the years there was not an hour of the day or night I did not look out at the country through the windows. The scenes were dark, mysterious, rural, and sparse. But when one approached L'viv or Kiev, the scene was alive with people.

One of the most surprising discoveries was the ability to communicate and understand by using or observing body language. That was all that was available. One day, in a meeting there was a discussion entirely in Ukrainian. Suddenly Kathy asked a question about the issue. The Ukrainians were shocked that she asked that question, as it was clearly relevant, even though she did not understand the language. We all became more observant. And we learned to speak slowly, distinctly, with simple vocabulary and short sentences and acting out what we meant when some of our employees were not conversant with English.

We had countless meetings in government and business offices. They were all nicely appointed. But there was never anything personal like a family picture. The offices of Ukrainian Wave were, after several years, in a building located next to the Railroad headquarters, and owned by the Railroad. We remodeled the floor for our offices which were quite nice. It was not so nice when it "rained", which happened when the toilets on the floor above our toilets overflowed. We were on the fourth story, and usually climbed the stairs. From the roof, outside our office, we had a wonderful view of the city. We also bought a small building in a downtown location for our sales office. Our technical equipment, the switch and computers were in the Ukrtelecom building, and we had three cell sites, including the one at Ukrtelecom.

We planned regular events for our employees. We would take them out for lunch or bring lunch to the office. Some of the employees also got used to working to midnight. Occasionally we invited them to the apartment when Kathy and Bill and I were there to cook. We took all the employees out to the country for an American Halloween Party. Everyone had to wear a costume. We had a cookout, with meat marinated in mayonnaise and grilled by O. Bilichenko. One of our gifts we often gave because it was so popular was a bottle of barbecue sauce. We had a treasure hunt, and bobbed for apples, including Bill and me, and even arranged a little dancing at the end. On this trip, Kathy was wearing an orthopedic boot, and on crutches. She had to negotiate her way on the slippery cobblestones to and from the office. We never saw anyone with such a boot or with crutches or anyone in a wheel chair. There were no accommodations for the handicapped.

One afternoon, the employees would not let me leave my office. Finally they said I should go to the conference room for a meeting. They were all there, including Bill and Kathy. It was a surprise, celebrating my 65th birthday! They gave me an antique silver pocket watch. On another occasion, we celebrated Kathy's birthday in a park with a cook-out.

One of the most interesting and telling experiences as to how the economy operated was opening a bank account. All over town, there were many places to exchange money. Just out from town, was the only bank that allowed foreigners to open a bank account. The bank was located on the second floor in a large room. Along the right side was a long wooden counter top, which extended at least 100 feet. Behind the counter there were desks, which extended the length of the counter. Each step in the process required the assistance of one of the people sitting at one of these desks. Jerry Merkelo and I each wanted to open an account and deposit $500. After about one hour into the process, a woman from an office told us that before we could complete the process we had to take a document to a notary and have it legalized. Back to town and to a notary, and then back to the bank. Several hours later, we made the deposit and received a bankbook. Every transaction included many documents. Every document required an official stamp. This meant that during the entire time we were there, the room reverberated with the sounds of wooden stamps on documents on the wooden countertop. It took us all morning to open our account.

One of Walter Bazarko's friends was the Dean of the L'viv University Law School. He later became a Judge of the National Constitutional Court. His intelligent and beautiful wife was also a Dean at the University. Once I was invited to meet the faculty of the Law School. I was invited several times to their dinner parties. They always insisted I tell my "tailor" joke. This was an action joke about a man who had a suit made for him, but one pant leg, one sleeve, and one lapel was too short. They loved the joke. At one party, the father of our hostess was there. During dinner I asked him if he had been in the Second World War. He left the table and then returned wearing his military coat, complete with many medals. He told us that he was one of the first Russian soldiers to enter Berlin.

He said he wrote his name on the Reichstag. After each dinner, their young daughter would perform ballet dancing for us. Many years later she attended Smith College at Northampton Massachusetts.

Over the years we visited most of the museums and many

historic buildings and restored Villages. Several times we went to St. George's cathedral (1744-1770) to hear Walter sing in the choir. But we never saw Walter sing in the choir. This was so because the church was always so crowded we had to stand outside and listen. On the edge of the center of the city is one of the oldest structures, the "powder tower" built in 1554. The round tower is now a monument to medieval fortifications.

The large Lychakivsky cemetery in L'viv is where many famous writers and poets are buried, including the famous writer and Poet, Ivan Franko who died in 1916. There are dozens of beautiful monuments, and many of the grave sites are candle lit, with many flowers. It is an historic and impressive cemetery. I visited it several times.

There are many restored houses and villages from the 17th and 18th centuries which we visited. Once I was driven about 40 miles to visit the restored Olesky Zamok (castle) that dated from the 14th to the 17th century. This historic and significant castle has been completely restored.

The considerable time and effort we spent in L'viv and Kiev was vital to starting this business. We were not just selling a product, then moving on. The business was located in L'viv. I hope this story demonstrates how important this effort was for the business and for us.

Part VIII: Final Summary

It is greatly appreciated that you took your time to read this story. Thank you! I would welcome your comments, suggestions or questions.

In deciding to start this business located in another country. Bill and I looked at our strengths and our limitations. But business, like life, never goes the way planned. Our potential assets were also our potential risks: A new independent country, with the desire to change from government control to a democracy and a market economy. The people and their government wanted the new future, but were also tied to their past. The greatest risk was not if, but when they could evolve from that past.

What emerged out of this complex cauldron was people became our greatest asset, our greatest strength and opportunity. People made the difference. Knowing, respecting and trusting each other requires an enormous effort and is an equal two way street. Just looking at Ukraine, people from the highest levels of government, from all segments of the community wanted to determine for themselves what it really meant to be an American business man. This undeniable fact should not encourage arrogance, but only humility. This story is filled with examples.

Let me expand on this theme. Our investors wanted to understand our business and financial plans. Equally important, they wanted to know if we were capable of delivering on these plans. But in Ukraine there was an additional element with strategic implications. They had limited experience and knowledge of American businessmen, but despite this, they had respect. I will give one additional example when I first gained this insight. During the time we were having an extremely difficult time with the government in Kiev, I was asked to meet with the Minister of Telecommunications for Ukraine. I immediately realized the importance of the meeting by the large number of people he asked to attend. He spoke first. He was a tall, slender man with a firm but calm voice. While he did not say this, he made it clear this was to be a discussion and that we were not there for them to beat up on us. He said as the Minister of Telecommunications, he knew the economy of Ukraine and he understood completely the finances of every telephone company in Ukraine. Since the economy of Ukraine could not support the number of telephone companies already doing business in Ukraine, there was no economic reason to start another one. Then he asked the question. "I believe Americans are good business men and they don't throw money away. So why are you doing this?"

Now that was not only a legitimate question, it was an excellent question. I will not detail my answer, primarily because I no longer remember it. I must have said things like we believe in the future growth of the Ukrainian economy as it develops into a market driven economy. We thought that if we built and managed a first rate company, we would beat the competition. If mergers and consolidations were necessary, we would do that.

Fortunately or unfortunately, there were not many American businessmen interested in starting a business in Ukraine, therefore, not many lessons to be learned from and therefore, he wanted us to succeed. He wanted to learn and determine if his support was justified. We had a detailed business plan, and an outstanding financial plan. I think he knew we were collaborative, problem solvers, and we worked hard. Especially important, he knew we had a strong commitment to making the business succeed. He gave us his support. His interest in Americans created s significant opportunity for us.

The most painful part of this story for me was the inability to build on their interest in Americans. This theme haunted me while I was there, and still does.

If there is such a thing as a classic friendship, then I was privileged to have that classic friendship with Bill. He was and still is, my best friend. As Chairman of both Hatwave and Ukrainian Wave, Bill has no equal. The glaring failures in telling this story would have been cured if Bill had been the co—author. This would have been a better story, if we had written it together. However, because we were so completely together, either one of us could write this story.

We saw firsthand the difficulties in transforming from a state-owned and controlled economy to a diverse market economy. Of course this will take many decades to accomplish. Yet, we also saw firsthand the determination and desire to make this transition.

When the window of opportunity is clear, it is too late. If a business is willing and able to take the long view, there are significant advantages of being on the ground floor. We saw foreign businesses doing just that.

Despite the difficulties, we were able to work with the people and the government both in L'viv and Kiev. I also met with people in other cities in Western Ukraine. We were positioned to expand to all of Western Ukraine. Even in the early years, I saw new brick houses being built all around the city. I have many pictures of these houses. I talked to many of the owners who wanted to know when they could take our service.

Epilogue

For five years, I have been working on this book, and a number of critical events have occurred.

On August 20th, 2012, we received information that the company had been sold again, and told that "Ukrainian Wave" switched off the last base station and switch. Now it is just a legal entity that doesn't provide any service." The engineers had a small beer party, "remembering all the good times that will never return." There was no public announcement of the change in ownership.

Finally, and most significantly, in June of 2013, we received a copy of the ruling in the Commercial Supreme Court of Ukraine, affirming the decisions of the Appellate Court denying our claims of the fraudulent theft of our companies. The litigation had been in process since 2008. This final ruling and all the prior Court rulings contained absolutely no substantive reasons why our proven evidence of fraudulent and illegal acts would not require invalidation of the theft of our companies. Consequently, the Shareholders of Hatwave will terminate Hatwave.

Acknowledgments

If life is a journey, then most important is to remember and THANK ALL OF YOU who made this journey possible. It is all of you who gave purpose and meaning to this journey. Many of the people we never knew before we began. Many were our close friends.

I have decided to thank our investors this way. You all, our friends, with unbelievable faith and loyalty took the greatest risk of all by giving us the financial support we so desperately needed. No words can adequately thank you, or to express how much that meant to us. You gave the most. We gave our best.

From Ohio Bell there was Gus Gustaferro, Executive Vice-President, who supported our early effort to start a wire line business in L'viv. My boss, Don Morrison, who was Vice President and General Counsel, always gave me his total support in all my endeavors with the Company. Dick Brown not only helped us at Ohio Bell, but also worked tirelessly with us in our office and in L'viv. During 1992, there were my two outstanding assistants, John Chandler and Barbara Sheers. Paul Karas was with me on my first trip to Russia and Ukraine. He gave me so much help and had the difficult task of trying to train this novice to do international business. There were many others from Ohio Bell, too numerous to mention.

We had other's support too. Maria Kaiser took outstanding pictures at our party in Cleveland to celebrate the launch of the business. From Peat Marwick, there was Jim Noteman,

Gerry Grant, and Dagnia Zeidlakis. From Jones Day, there was Karl Herold, who also read the manuscript and wrote a beautiful letter to us which is part of the Preface, David Watson, Doug Whipple, and Yuri Zaichuck, and his brother Oleg an attorney in Kiev. From the EBRD were Mark Tomlinson and Carl Gage, Ron Freeman, and Dellarosia, the Chairman of the EBRD Board of Directors. Jerry Merkelo was our first Managing Director, who helped us through the process, and recruited an outstanding staff. Andy Fedesky, whose father founded the significant Ukrainian Museum in Cleveland, found our translator, Anatole Siry. Anatole worked with us during the critical period of the shipping of equipment and preparation of the customs documents, as well as translations of some of the legal documents into Ukrainian. He was able to help us understand the importance of original documents required in Ukraine.

All through this saga, Lou Masterson provided space and office support for all our efforts in Cleveland. Our accountant and tax consultant, Ray Klinc, provided constant invaluable service. Neither of these two great supporters was paid for their services. Three friends, going back to the beginning of my legal career, Harry Lehman, Jim Friedman, and Mike Honohan, gave us valuable advice and legal assistance.

To those who read drafts of the manuscript and made valuable suggestions: As already quoted in the preface, Lute Harmon Sr. who not only read the manuscript, he wrote a kind full page review in his July,2010 issue of "Inside Business." What he wrote was meaningful to us, and totally unexpected; also, Glen and Mim Blair, Margret Cowin, Judy McMillan, Charlie and Alice Butts, Andy Fedesky, Charlie

Marotta, Gerry Gordon, and Andy Bihun.

From Washington D.C. was our consultant, V. Garbar, and Andy Bihun, the Commercial Attaché to Ukraine, and Ambassador Miller.

From Hughes there were Arunas Slekys and others including Jud Kenny and Wheeler Ramey and Carl from Alcatel. From OTE, there was Manolis Georgakakis, Vassilios Maglaras, Yanis Kaligirou, and Costas Petrides.

From Ukraine, there were two ministers of Telecommunications, and Ms. Phillipova; Rubin.Belieav, an employee of the US embassy. Nick and Sveta Sergienko were our dear friends in Kiev. There were and are others in Ukraine with whom Kathy kept in email contact that were enormously helpful, but shall remain anonymous.

The Ukrainian employees of UW included O. Bilichenko and V. Kalynyuk, both Managing Directors, the engineers Max, Slavko, Ihor and Roman. Among the many very helpful employees were Yarema, Ihor, Hanna, Ivan, and Dachko. Ukrainian Board members included S. Dembicky, O. Kryskiv, H. Kirpa, and P. Hnatenko, V.Orlenko, and S. Gonchar from the State Property Fund. Two mayors of L'viv were always helpful and so were many other friends too numerous to mention.

To the children of both partners: Bill, Steve, and David Schlageter, and Vincent, Sarah, and Heather Stoll. You all were supportive and terrific.

Most importantly, I'd like to thank the great sacrifices made by Linda Schlageter, Lydia Bazarko, and my wife Kathy. Kathy worked so hard for us in the business. During the past five years spent on writing this book, every word on every page was entered on the computer by Kathy. What a unique privilege it was for a husband to gain such admiration and respect for his wife.

Finally and for obvious reasons I saved the dedication of this book to the end. To Bill Schlageter, my best friend and business partner, this book is dedicated to you. What a rare privilege in life to have such a classic friendship.

###